Lecture Notes in Bioengineering

Lecture Notes in Bioengineering (LNBE) publishes the latest developments in bioengineering. It covers a wide range of topics, including (but not limited to):

- Bio-inspired Technology & Biomimetics
- Biosensors
- Bionanomaterials
- Biomedical Instrumentation
- Biological Signal Processing
- Medical Robotics and Assistive Technology
- Computational Medicine, Computational Pharmacology and Computational Biology
- Personalized Medicine
- Data Analysis in Bioengineering
- Neuroengineering
- Bioengineering Ethics

Original research reported in proceedings and edited books are at the core of LNBE. Monographs presenting cutting-edge findings, new perspectives on classical fields or reviewing the state-of-the art in a certain subfield of bioengineering may exceptionally be considered for publication. Alternatively, they may be redirected to more specific book series. The series' target audience includes advanced level students, researchers, and industry professionals working at the forefront of their fields.

Indexed by SCOPUS, INSPEC, zbMATH, SCImago.

Magdalena Kobielarz · Adam Wittek ·
Martyn P. Nash · Poul Nielsen · Anju R. Babu ·
Karol Miller
Editors

Computational Biomechanics for Medicine

Progress in Research and Applications

Springer

Editors
Magdalena Kobielarz
Faculty of Mechanical Engineering
Wrocław University of Science
and Technology
Wrocław, Poland

Martyn P. Nash
Auckland Bioengineering Institute
University of Auckland
Auckland, New Zealand

Anju R. Babu
Department of Biotechnology and Medical
National Institute of Technology Rourkela
Rourkela, Odisha, India

Adam Wittek
Intelligent Systems for Medicine Laboratory
The University of Western Australia
Perth, WA, Australia

Poul Nielsen
Auckland Bioengineering Institute
University of Auckland
Auckland, New Zealand

Karol Miller
Intelligent Systems for Medicine Laboratory
The University of Western Australia
Perth, WA, Australia

ISSN 2195-271X ISSN 2195-2728 (electronic)
Lecture Notes in Bioengineering
ISBN 978-3-031-94127-6 ISBN 978-3-031-94128-3 (eBook)
https://doi.org/10.1007/978-3-031-94128-3

This Springer imprint is published by the registered company Springer Nature Switzerland AG
The registered company address is: Gewerbestrasse 11, 6330 Cham, Switzerland

Preface

Connections between computing and engineering technologies and medicine are still challenging. Rapid scientific and application progress in both disciplines is more than impressive. Facing these unique times, the Workshop of Computational Biomechanical for Medicine XIX creates opportunities to discuss new approaches, solutions, methods, and theories. This is a step to support the mutual permeation of these two disciplines, focusing on the latest challenges and achievements. Hence, the nineteenth workshop was held under the leading title Progress in Research and Applications.

The Workshop of Computational Biomechanical for Medicine XIX was held in conjunction with the 27th International Conference on Medical Image Computing and Computer Assisted Intervention (MICCAI 2024) on October 06, 2024, in Morocco. Like every year since the fourth workshop organised in 2009 in London (United Kingdom), the result of this unique scientific meeting of researchers is the book Computational Biomechanical for Medicine-Progress in Research and Applications. Manuscripts (8-15 pages) submitted for the workshop passed a rigorous evaluation and at least two independent reviews were performed. Twelve of the best manuscripts are included in this volume. The authors are researchers from Australia, Germany, France, New Zealand, Poland, and Switzerland.

Magdalena Kobielarz
Adam Wittek
Martyn P. Nash
Poul Nielsen
Anju R. Babu
Karol Miller

Organization

Scientific Committee

Reviewers

Ghulam Mubashar Hassan	Department of Computer Science and Software Engineering, the University of Western Australia, Perth, Western Australia, Australia
Marek Pawlikowski	Warsaw University of Technology, Warsaw, Poland
Mariusz Ptak	Wrocław University of Science and Technology Wrocław, Poland
Samarth S. Raut	Indian Institute of Technology, Delhi, India
Roy Sitikantha	Indian Institute of Technology, Delhi, India
Pavel Skacel	Brno University of Technology, Brno, Czech Republic
Kamil Sybilski	Military University of Technology, Warsaw, Poland
Ewelina Świątek-Najwer	Wrocław University of Science and Technology Wrocław, Poland
Brigitta Tóth	Budapest University of Technology and EconomicsBudapest, Hungary
Andreas Wittek	Frankfurt University of Applied Sciences, Frankfurt, Germany
Adam Wittek	Intelligent Systems for Medicine Laboratory, the University of Western Australia, Perth, Western Australia, Australia
Wiktoria Wojnicz	Gdańsk University of Technology, Gdańsk, Poland
Wojciech Wolański	Silesian University of Technology, Gliwice, Poland
Benjamin F. Zwick	Intelligent Systems for Medicine Laboratory, the University of Western Australia, Perth, Western Australia, Australia

Contents

Challenges and Bottlenecks for a Clinical Use of the Biomechanical Digital Twins

Yohan Payan(✉)

Univ. Grenoble Alpes, CNRS, UMR 5525, VetAgro Sup, Grenoble INP, TIMC, 38000 Grenoble, France
Yohan.Payan@univ-grenoble-alpes.fr

Abstract. This invited talk will address the design of biomechanical models of living organs, in the context of computer assisted medical interventions. Many models have been proposed by research labs around the world but very few have been transformed into medical devices that are used in the clinical routine. Two main bottlenecks seem to explain this and will be discussed during the talk. First, it is known that each organ model should be patient-specific, both in terms of geometry of the organ and in terms of constitutive parameters of tissues included in the organ model. Such a personalization of the models can be extremely time consuming and therefore not compatible with the clinical constraints. Secondly, each patient-specific organ model has to be used by the clinician in an interactive way, with simulations that should not require more than a couple of seconds or minutes to compute. Here again, this can be very challenging when complex hyper-elastic models are used.

M. Kobielarz et al. (Eds.): MICCAI 2024, LNBE, p. 1, 2025.
https://doi.org/10.1007/978-3-031-94128-3_1

Validation of 4D Ultrasound Strain Imaging of the Aortic Wall

Christopher Blase[1,2(✉)], Wojciech Derwich[3], Achim Hegner[2], Thomas Schmitz-Rixen[3], Kyriakos Oikonomou[3], Armin Huß[1], and Andreas Wittek[1]

[1] Frankfurt University of Applied Sciences, 60318 Frankfurt am Main, Germany
cblase@fb2.fra-uas.de
[2] Goethe University, 60629 Frankfurt am Main, Germany
[3] Goethe University Hospital, 60590 Frankfurt am Main, Germany

Abstract. Time resolved 3D ultrasound (4D ultrasound) is a promising imaging modality to measure geometry and deformation of cardiac structures, because of its high temporal resolution. Previously, we have established 4D ultrasound with speckle tracking to measure the temporally and spatially resolved deformation of the aorta and aortic aneurysms. The measurement accuracy of geometry and deformation measurement by 4D ultrasound was determined by comparison to optical imaging in an in vitro inflation/extension test of porcine aorta, subjected to cyclic loading (five load cases, ranging from pulse pressures of 7 to 90 mmHg and axial prestretches of 1.31 to 1.61). The resulting diameter, diameter change, local displacement and circumferential strain were calculated from 4D ultrasound and optical imaging, respectively. The diameter measurement showed a systematic error of up to –0.55 [0.96] mm (median [IQR]), whereas it was negligible for measures of deformation (up to –0.04 [0.21] mm). The random error of single ultrasound-based measurements was higher for measurements of geometry (95% confidence interval up to ~ 2.5 mm) than for measurements of deformation (95% c.i. up to ~ 0.5 mm). The random error did not scale with the measured deformation amplitude, resulting in relative random errors up to 100% of the measured values for small deformations. The random error could be significantly reduced by averaging of several evaluations of the same ultrasound data set. The reproducibility was assessed by the intraclass correlation coefficient and found to be excellent with values ≥ 0.90. 4D ultrasound-based measurements provide accurate measures of deformation for large blood vessels, with a considerable random error in the case of small deformations, which can be reduced by averaging results from repeated evaluations.

Keywords: 3D ultrasound · strain imaging · validation · uncertainty

Abbreviations

4D US - time resolved 3D ultrasound imaging,
ACP - advanced cardiac package,
DIC - digital image correlation,

M. Kobielarz et al. (Eds.): MICCAI 2024, LNBE, pp. 2–15, 2025.
https://doi.org/10.1007/978-3-031-94128-3_2

ICC - intraclass correlation coefficient,
IQR - interquartile range,
RF - radio frequency,
US - ultrasound.

1 Introduction[1]

4D ultrasound (4D US)-based measurement and reconstruction of the 3D geometries of the heart and of large blood vessels has been validated previously in vivo and in vitro [1–7]. However, none of these previous studies was transferable to the use of 4D US for the measurement of aortic and aneurysmal wall motion that was established by our group [7–10]. Originally, the used device, Artida (Toshiba Medical Systems, Otawara, Japan) was designed and validated for the assessment of left ventricular wall motion [2, 4]. In the work of our group, it was used to measure the motion of the aortic wall, which is a far thinner structure undergoing smaller deformations, compared to the left ventricular wall. Thus, validation of 4D US measurement for this specific application seemed to be necessary.

Several of the previous studies refer to deformation parameters of the myocardium that are indicative for left ventricular function [1, 2, 4, 6]. Other studies have used other ultrasound devices, post-processing algorithms and routines than the ones that were applied within our work [3, 11, 12]. Finally, an earlier validation study of the application of the Toshiba Artida to aneurysm-like structures [5] did use silicone phantoms instead of aortic soft tissue and performed tests in distilled water at room temperature. Both, specimen and medium, showed lowered speeds of sound compared to soft tissue which might have impaired the results of the validation.

The objectives of this validation were to quantify the uncertainty and the reproducibility of 4D US full-field displacement measurement of aortic wall tissue and derived strain fields (cf. [13, 14] for methods and concepts of validation of measurement techniques). The validation addressed the measurement by use of the Toshiba Artida device itself, combined with post-processing with the customized Advanced Cardiac Package (ACP, Toshiba Medical Systems, Otawara, Japan). Possible sources of uncertainty and impaired reproducibility of 4D US measurements were, firstly, inherent in its physical principle, e.g. the wavelength of the ultrasound wave might have restricted spatial resolution of the measurement. Secondly, the observer dependency of the measurement (hand-held probe) and of the semi-automatic segmentation of the vessel wall, which defined the initial shape and the region of interest for the speckle tracking algorithm, were possible sources of uncertainty. Finally, the parameters of the measurement such as the forming of the used wave packet, focusing and beamforming of the RF signal by control of the phased-array matrix probe and the post-processing of the RF data including the speckle tracking algorithm were proprietary technology and unknown to the authors. It was the aim of the presented study to quantify the resulting lumped effect of different sources of uncertainty on the measurement. To identify and weight the influence of different

[1] The current chapter is based on chapter 4 of the PhD thesis of Andreas Wittek [10].

sources of uncertainty was beyond the scope of this work. The uncertainty and reproducibility of a single 4D US measurement was to be quantified because measurements for diagnostic purposes are taken only once, in clinical practice.

Measurement uncertainty was assessed as agreement of the 4D US measurement with optical reference measurements. The reproducibility of the 4D US measurement was assessed in terms of its test-retest reliability, i.e. the degree of reproducibility of a measurement under identical circumstances (subject, device, observer) [15, 16]. For this purpose, an Analysis of Variances of repeated ultrasound measurements under identical conditions was performed.

2 Methods

2.1 Sample and Experimental Set up

Sample Preparation. Thoracic aortas of young pigs (10–12 months) were obtained from a local slaughterhouse and stored in phosphate buffered saline (pH 7.2) with 10% (v/v) dimethyl sulfoxide at –20 °C. A straight tubular specimen of porcine aorta with dimensions of 15 mm x 130 mm (luminal diameter x length in its load-free configuration) was carefully separated from attached tissue and side branches and fixed to custom mounts. An irregular speckle pattern was applied to the specimen surface for the purpose of optical deformation measurement by means of digital image correlation. after the experiments, ring-shaped cross sections were cut from both ends of specimen and were photographed. Using ImageJ v. 1.48 [17] the average load-free wall thickness of the specimen was determined to be $s_0 = 1.93$ mm.

Experimental Set up and Load Cases. Experiments were performed using a custom-built inflation-extension set up (Fig. 1) that allowed for a constant axial extension (axial prestretch) and dynamic inflation (cyclic pressure change) of the sample. Deformation was measured in parallel optically by two 8-bit CMOS cameras (1.3 Megapixels) and by 4D US. The reduced axial force (Fig. 1, device 6) and the transmural pressure (Fig. 1, device 4) were measured. Image and sensor data acquisition was controlled by custom LabVIEW programs (LabVIEW 2011, National Instruments, Austin, TX, USA) on a PC. The specimen was mounted in a cuboid water bath with heated saline (37 °C, 0.9% (w/v) NaCl) in order to provide acoustic properties of the medium close to the in vivo conditions, in particular, the sound velocity of 1540 m/s in human abdominal soft tissue [19]. Five different combinations of axial prestretch, inner base pressure and cyclic pulse pressure were applied to the same specimen that provided characteristically different deformed reference geometries (diameter, wall thickness) and cyclic deformation (Table 1) with a frequency of 0.8 Hz.

4D Ultrasound Imaging. By 4D US, full field deformation measurements of a 5–8 cm segment of the specimen were performed using a customized commercial Artida real-time 3D US device that was equipped with a 3D transthoracic probe (Toshiba, PST-25SX, 1–4 MHz phased array matrix transducer). All measurements were taken by the same experienced observer. The probe was fixed by a custom-made mount, reducing the observer influence. Combined with speckle tracking analysis using a commercial algorithm (ACP), this imaging technique provided 3D position vector fields of up to 1188

Table 1. Parameters of the applied load cases and resulting loaded reference configuration and cyclic deformation of the porcine aortic specimen as obtained from optical reference measurements. Ref. wt ≡ wall thickness in the deformed reference configuration, circ. ≡ circumferential.

	C01	C02	C04	C07	C08
Max./min. p in mmHg	113/26	105/83	128/45	101/94	158/68
Pulse pressure Δp in mmHg	87	22	83	7	90
Axial prestretch $\lambda_{1,r}$	1.31	1.31	1.42	1.61	1.61
Circ. Prestretch $\lambda_{2,r}$	0.95	1.06	0.94	1.07	1.00
Deformed ref. wt in mm	1.6	1.4	1.4	1.3	1.2
Reference diameter in mm at p_{min}	17.6	20.2	18.0	20.3	19.0
Cyclic diameter change in mm	3.6	0.87	3.23	0.23	2.82

measuring points on the luminal surface of the specimen for each of about 40 temporal frames throughout a loading cycle, corresponding to approximately 33 frames/s. These data describe the trajectories of discrete material points corresponding to surface areas of about 5 mm^2 [10, 18]. The x-axis of the US probe coordinate system corresponded to the longitudinal direction of the specimen. The y-axis corresponded to the dominant propagation direction of the ultrasonic waves and the z-axis was orthogonal to the propagation direction (cf. Figs 1 and 2). The ultrasound probe coordinate system was used in further analysis of all data and optical data were transformed accordingly (cf. section on optical imaging).

Two successive 4D US measurements of each load case were taken. Post-processing of each of the 10 4D US data sets was performed 10 times by one experienced observer, including the observer dependent semi-automatic segmentation of the aortic wall as region of interest and the application of the proprietary speckle tracking algorithm. In total, 20 4D US evaluations were available for each load case.

Optical Imaging and Post-Processing of Optical Deformation Data. Optically, the deformation of the specimen's outer surface was measured by two 8-bit CMOS cameras with a resolution of 1280 x 1024 pixels corresponding to a pixel size on the specimen surface of approx. 0.1 mm. Temporal resolution of optical measurements was about 25 images/s. Optically visible steel reference axes of coordinates (Fig. 2b) were used for the alignment of the cameras in line with, and orthogonal to the y-axis of the US probe coordinate system (Fig. 2). A reference for a common origin of the x-axis was defined by identifying the position of the tip of a plummet from US and optical measurements, which are shown in Fig. 2a and b. This spatial arrangement allowed for the separate assessment and comparison of the accuracy of the 4D US measurement in propagation direction (y-axis) of the US waves and orthogonal to the propagation direction (z-axis) (Fig. 2c).

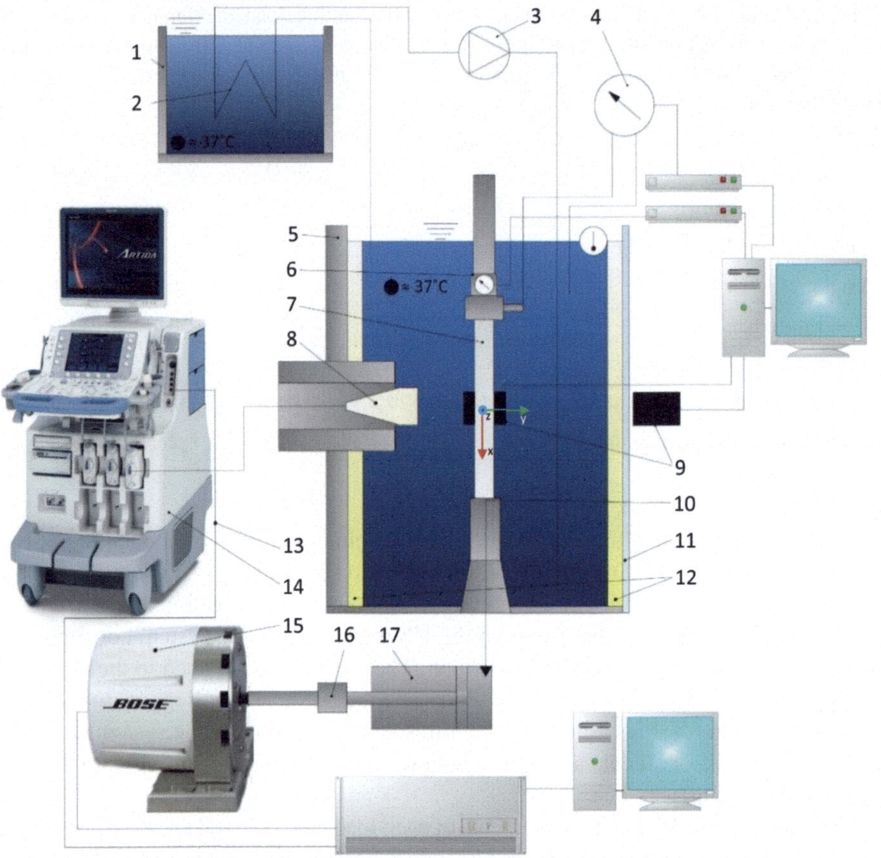

Fig. 1. Inflation-extension testing device equipped with 4D US and optical measuring devices: (1) heated bath, (2) heat exchanger, (3) rotary pump, (4) differential pressure sensor, (5) tank (front), (6) axial force sensor, (7) sample (porcine aorta), (8) US probe fixed by custom made mount, (9) optical cameras (8-bit CMOS, 1.3 Megapixels with 8 mm focal length objectives), (10) pressuring tube, (11) tank, (12) illumination (LED), (13) trigger, (14) 4D US device, (15) linear motor (Electroforce LM1, Bose Corp., USA), (16) coupling, (17) pressure cylinder. US probe coordinate system is indicated in the center of the sample (cf. Fig. 2).

Optical reference values for the evaluation of 4D US measurement uncertainty were obtained for each experiment as the means of three subsequent measurements in order to minimize the measurement uncertainty. Alignment and spatial calibration of the optical measurements was done separately for each camera by using a custom target. Local reference diameter and diameter change of the specimen, spatially resolved along the x-axis of the optical coordinate system, were determined in the xy-plane applying an in-house threshold-based edge detection method.

Temporally resolved 2D displacement fields in the xy- and the xz-plane, respectively, were determined using a modified digital image correlation (DIC) code proposed by [20]. Both, edge detection and DIC, were coded using MATLAB (The Mathworks, Inc.,

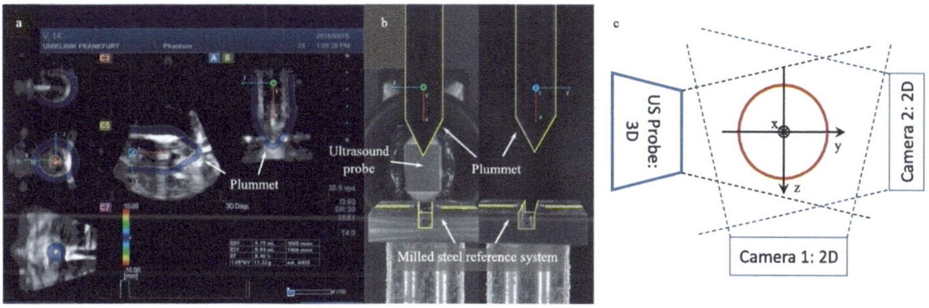

Fig. 2. (a) US and (b) optical reference measurements of a plummet as basis of a common coordinate origin in x-direction. The steel reference coordinate axes (b, bottom) were used for the alignment of the CMOS cameras with the US probe coordinate system. Plummet and steel reference axes are highlighted in yellow for better recognizability. (c) Set-up of US probe and two CMOS cameras for simultaneous measurements.

Natick, Massachusetts, USA). A subset size of 31 was chosen for all DIC evaluations. A threshold of the normalized 2D cross-correlation coefficient of 0.7 was used to exclude non-correlating regions and avoid artificial displacement values.

2.2 Quantification of the Measurement Uncertainty: Agreement of 4D US and Optical Measurements

Optical measurements provided 2D data of the outer diameter in the *xy*-plane and two mutually orthogonal 2D displacement fields on the outer surface of the wall. In contrast, 4D US provided 3D position vector fields on the inner or luminal surface of the wall. In order to compare both, firstly, 4D US data were corrected for the deformed wall thickness. The average deformed wall thickness was computed for each of the five load cases using the incompressibility constraint and applied as radial offset to each position vector describing a material point of the specimen surface. Comparison of diameter values used the outer diameter, only.

In order to compare 4D US data to the available optical diameter measurements in the *xy*-plane, the measurement points with the maximum and minimum *y*-values, respectively, were identified from the US field data. Then, the local diameter of the specimen was computed as difference of maximum and minimum *y*-value of corresponding measurement points along the longitudinal specimen (*x*-)axis for each temporal frame. Local diameter change was computed for each *x*-position with regard to the temporal reference frame showing the minimum mean diameter. Finally, local circumferential strain was obtained as ratio of local diameter change and reference diameter.

Because of a missing common temporal reference signal, US and optical data had to be synchronized for the purpose of comparison. The temporal delay τ of the optical measurement with regard to the US data was determined by means of the cross-correlation (SPEARMAN‘s ρ) between US and optical diameter change data. The optical diameter, diameter change and strain data were shifted accordingly.

The deviation of diameter, diameter change, and circumferential strain measured by US from the corresponding optical measurements of the same load case was evaluated for *local* values along the length of the specimen and for *average* values, i.e. the arithmetical mean of the local values along the sample length.

For the comparison of displacement field data, the optical data were synchronized with the US data applying the method described above. Two-dimensional motion functions for the optically visible material points in the xy- and the xz-plane were derived from the 3D US motion function using the x- and y-components and the x- and z-components of the selected material points in their current configurations, respectively. Displacement fields were determined with regard to the reference configuration. Displacement field data were not compared for all frames, but only the displacement fields showing the maximum mean displacement with regard to the reference configuration.

All repeated measurements and evaluations of the US data were compared to the identical optical reference measurement, separately. Results of the comparison were given as median [IQR] where IQR is the interquartile range.

2.3 Quantification of the Reproducibility: Test-Retest Reliability

The maximum values of the (i) mean and local diameter, (ii) mean and local diameter change and (iii) mean and local circumferential strain, that were obtained throughout the load cycle from repeated US measurement (2 times per load case) and post-processing (10 times per measurement) of the specimen's deformation under 5 different load cases were analysed. No fixed effect or systematic error was included in the model because the repeated measurements and post-processing steps were carried out under identical conditions (operator, US device and settings, speckle tracking algorithm). Since data were non-normally distributed, the KRUSKAL- WALLIS test was used to test whether the distributions of the results from 20 repeated measurements and post-processing of the same load case were significantly different between 5 load cases. In order to assess the test-retest reliability of a single measurement the intraclass correlation coefficient ICC(1,1) [15, 21, 22] was determined using Wolfram Mathematica 10.

3 Results

3.1 Agreement of 4D US and Optical Measurements

Table 2 gives an overview of the deviations of the 4D US measurements of average and local diameter, diameter change, circumferential strain in the xy-plane and local displacements in y- and z-direction from the corresponding optical reference measurement as obtained from the evaluation of all load cases pooled.

Average and local diameter change, strain and local displacement measurement in y- and z-direction did not show any systematic error. The average and the local diameter values exhibited a systematic deviation: US systematically underestimated the diameter of the specimen by more than 0.5 mm. Figure 3 illustrates the differences in the systematic and random error of the measurements of diameter and diameter change.

Average and local diameter measurements showed the by far largest random errors of IQR = 0.79 mm and IQR = 0.96 mm, respectively, corresponding to the 95% confidence intervals [–1.58 mm, 0.70 mm] for average values and [–1.78 mm, 0.79 mm] for local

Table 2. Absolute deviations of the results of the 4D US measurements with respect to optical reference measurements. Results are given as median [IQR].

	Average	Local
Deviation of outer diameter in *xy*-plane and derived quantities		
Deviation of outer diameter in mm	−0.52[0.79]	−0.55[0.96]
Deviation of outer diameter change in mm	−0.01[0.16]	0.00[0.22]
Deviation of circumferential strain in %	−0.01[0.98]	0.04[1.21]
Deviation of displacements		
Deviation of u_y in *xy*-plane in mm		−0.04[0.21]
Deviation of u_z in *xz*-plane in mm		0.01[0.18]

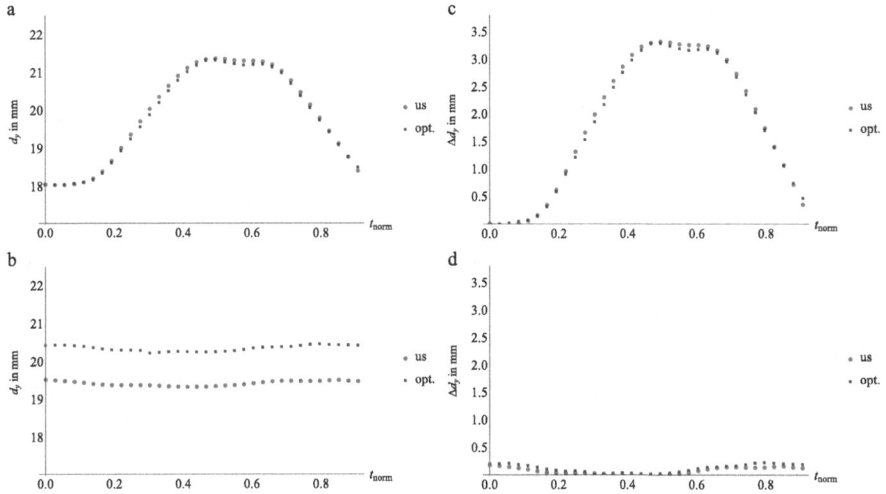

Fig. 3. Spatially averaged outer diameter (a, b) and diameter change (c, d) over one load cycle for load cases C04 (a, c) and C07 (b, d). The results of the opticalmeasurement (opt, blue) and of the 1st evaluation of the 1st US measurement (us, red), are shown.

values. In contrast to this, a smaller random error of IQR = 0.16 mm or a 95% confidence interval of [–0.32 mm, 0.27 mm] was found for average diameter change.

Compared to measures of average deformation along the length axis of the sample, local deformation measures exhibited slightly larger random error values, characterized by similar IQR ≈ 0.2 mm and 95% confidence intervals of [–0.41 mm, 0.31 mm] for diameter change in the xy-plane, [–0.47 mm, 0.49 mm] for local displacements in the xy-plane and [–0.32 mm, 0.52 mm] for local displacements in the xz-plane. Figure 4 gives a comparison of single 4D US full-field displacement measurements (not the mean of repeated measurements and evaluations) with the optical reference measurements for two load cases. Therefore, it illustrates the larger random error of a single 4D US measurement, not the negligible systematic error of repeated measurements and evaluations. Though this is an exemplary representation of the validation results, only, it is worth noting that the displacement fields resulting from the load case with the smallest pulse pressure of 7 mmHg, only (cf. Table 1, C07, Fig. 4 b, d), show the most considerable qualitative deviations.

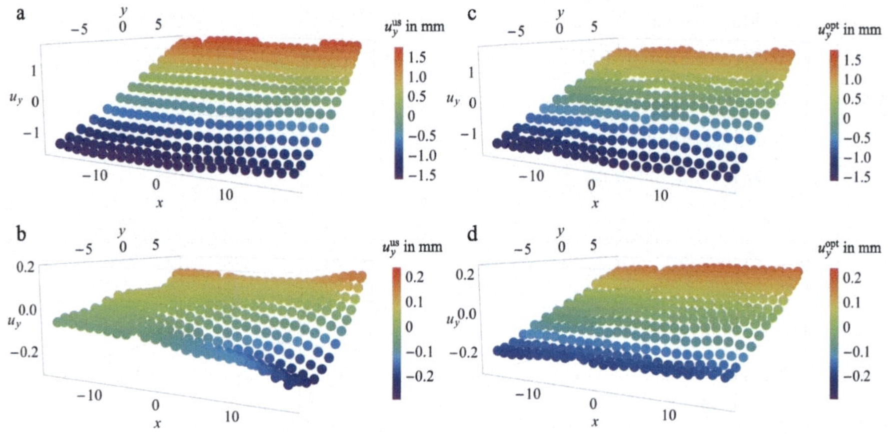

Fig. 4. Illustrative comparison of the results of local y-displacement (u_y) in the *xy*-plane as obtained by 4D US (left column: a, b) and optically (right column: c, d). Exemplarily, the results of single US measurements and evaluations, not the means of repeated measurements and evaluations, are shown for the load cases C04 (a, c) and C07 (b, d).

The random error of local strain measurement (IQR = 1.21%, [−2.1%, 2.0%]) was slightly larger than the random error of averaged strain (IQR = 0.98%, [−1.6%, 1.6%]).

3.2 Reproducibility of 4D US Measurements

Test-Retest Reliability and Intraclass Correlation Coefficient. An overview of the results for repeated measurements and evaluations of the five different load cases is given in Table 3. Repeated determination of average and local diameter showed a clearly larger IQR of 0.32 mm and 0.41 mm, respectively, compared to the variation of the repeated determination of average and local diameter change, which was characterized by an IQR of 0.12 mm and 0.05 mm, respectively. For repeated measurements of average and local strain, IQR values of 1.0% and of 0.6% were determined, respectively.

The Kruskal-Wallis test for non-parametric distributions proofed that the median values of the results of repeated tests were significantly different ($p < 0.001$) between the examined load cases for all deformation parameters. For all examined deformation parameters ICC (1,1) values ≥ 0.90 were observed indicating excellent test-retest reliability of the 4D US displacement and strain measurement (including post-processing). The capacity of 4D US to characteristically distinguish the five load cases that were examined in this study is illustrated in Figs. 5a and b with regard to local diameter and circumferential strain.

Table 3. Results for the test-retest reliability of the measurement of local outer diameter, diameter change and circumferential strain along the sample. The distributions of 20 repeated evaluations of the 4D US measurement of five different load cases (cf. Table 1) were compared. H-Test refers to the Kruskal-Wallis test, and ICC (1,1) to the intraclass correlation coefficient.

		IQR	H-Test, p	ICC(1,1)
Outer diameter in mm	averaged	0.32	<0.001	0.92
	local	0.41	<0.001	0.90
Diameter change in mm	averaged	0.12	<0.001	0.96
	local	0.05	<0.001	0.97
Circumferential strain in %	averaged	1.0	<0.001	0.96
	local	0.6	<0.001	0.97

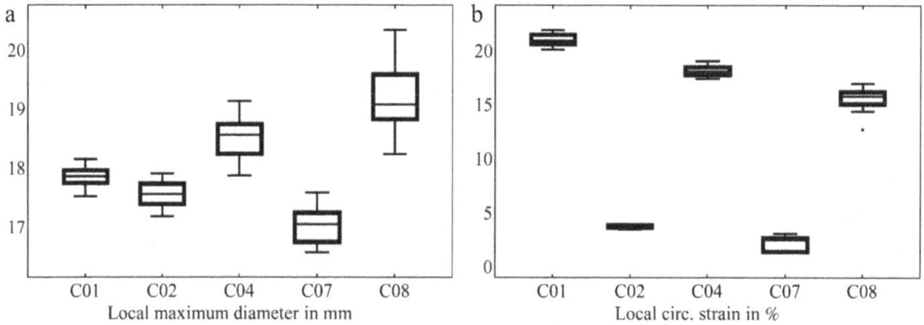

Fig. 5. Results for the test-retest reliability of the measurement of local outer diameters (a) and local circumferential strain (b) along the sample. The distributions of 20 repeated evaluations of the 4D US measurement of five different loading conditions (C01, C02, C04, C07, C08) are compared. Boxes show the median and range of *Q1* to *Q3*, whiskers extend to minimum and maximum values, respectively (outliers are indicated by filled circles). Kruskal-Wallis test $p < 0.001$, ICC (1,1) = 0.90 (a) and 0.97 (b), respectively.

4 Discussion

In this study, the uncertainty and the test-retest reliability of 4D US with speckle tracking were assessed. The uncertainty was determined by comparison to optical reference measurements, which, however, were subject to uncertainty themselves. The uncertainty of the optical measurements was mainly determined by the pixel size of 0.1 mm. Their random error was minimized by determining the reference values as the means of three subsequent measurements. It is worth noting and an indication of the robustness of the results of this study that the agreement of 4D US measurements with optical measurements were of the same size for diameter change and local displacement (systematic error: median $\approx$ 0.0 mm, random error: IQR $\approx$ 0.2 mm): diameter change and local displacement were derived independently and based on different principles from primary optical image data.

In this study, for the measurement of the diameter a systematic underestimation of about 0.5 mm to 0.6 mm and a random error of about 2.5 mm were observed. In contrast, deformation measures such as diameter change, circumferential strain and displacement did not show any systematic error. Moreover, considerably smaller absolute random error was observed for these relative measurements of geometry change: 95% confidence intervals between 0.6 mm and 1.0 mm (corresponding to 1.96 sd = $\pm$ 0.3 mm... $\pm$ 0.5 mm) were determined for diameter change and displacement. A possible explanation for this difference in the accuracy of the measurement of geometrical dimensions and of deformation could be the fact that the geometry or shape of the measured structure was defined by manual segmentation of the aortic wall in two longitudinal sections of the already post-processed and visualized data. DICOM image data on which the manual segmentation of the shape was performed typically exhibited a voxel size between 0.4 and 0.6 mm. That means that a deviation of 0.5 mm to 1.0 mm from the true shape could not be controlled or avoided in the process of manual segmentation. In contrast, the measurement of deformation or displacement was performed automatically by the speckle tracking algorithm that was implemented in the Toshiba post-processing software (ACP). Starting from the manually segmented region, cubic template volumes of approximately $10 \times 10 \times 10$ mm ^{3} were defined automatically by the software. Displacement of the center points of these template volumes in subsequent time steps was estimated by means of pattern recognition on the template volumes. Consequently, the uncertainty of the spatial location of these template volumes due to variation of manual segmentation and the observed systematic error of shape measurement were in the range of the voxel size of the DICOM data, i.e. $\approx$ 0.5...1.0 mm. This means that the template volume with at size of 1000 mm^3 and the spatial pattern which was the basis of motion estimation by the speckle tracking algorithm were substantially unchanged by and, consequently, estimated displacements were widely independent of variations due to the uncertainty of the manual segmentation. Another aspect worth recognizing in this context is, that speckle tracking using the ACP was not performed by digital image correlation on already post-processed DICOM data, but by means of cross-correlation on the original radio frequency data, which should contribute to a high accuracy and reliability of displacement measurements [23]. Still, because of the size of the measured structure, i.e. the specimen diameter of about 20 mm, the systematic (–5.8% to 0.2%) and random relative error (1.7% to 5.2%) of diameter measurement was small.

The observed uncertainty of 4D US displacement measurement in y- and in z-direction were of the same size (systematic error $\approx$ 0.0 mm, random error $\approx$ 0.2 mm), although the y-direction corresponded to the propagation direction of the RF signal, i.e. the accuracy depended on the wavelength and corresponding frequency, type of wave packet and beamforming of this signal. In contrast, the measurement accuracy in z-direction, i.e. orthogonal to the propagation direction, depended on the design and sequential control of the matrix array of the US probe. The fact, that equal systematic and relative uncertainty was observed in both directions indicates that the uncertainty of highly resolved 4D US displacement measurements is direction independent.

The chosen load cases consisted of two distinct groups of experiments: three experiments with displacements > 1 mm ("large deformations") and two load cases with displacements < 0.5 mm ("small deformations"). Large deformation load cases showed negligible systematic error of diameter change and absolute random error values of IQR = 0.08…0.21 mm resulting in relative random error of IQR = 2.5…7.5%. In contrast, systematic deviation of diameter change measurement between –0.13 mm and + 0.13 mm and larger absolute random error of IQR = 0.21…0.25 mm were observed for the small deformation load cases, indicating that the absolute error of the speckle tracking algorithm may require a minimum deformation to work properly. In combination with the smaller size of the measured value, the larger absolute error values for the small deformation load cases resulted in large systematic relative errors between –15.1% and + 56.7% and even larger relative random errors up to IQR = 94.8%.

Table 4. Literature values of 3D and 4D US based measurements of geometry and deformation.

Measurement	Reference measure	Relative deviation (systematic/random)	Reference
Geometry Aortic root volume / 3D US	CT	3.5% / 50%	[3]
Geometry left ventricle (LV) volume	MRI	-3.3% / 9.2% -3.4% / 6.8%	[1]
Geometry abdominal aortic aneurysm	CT	Haussdorff distance (Median / IQR): 4.8 mm / 3.1mm 13.9mm / 9.9 mm	[11]
Deformation LV ejection fraction	MRI	0.2% / 6.2%	[1]
Deformation LV wall strain (LS/CS/RS)	Sonomicrometry	LS: 1% / 11%, CS: -3% / 13%, RS: 3% / 14%	[2]
Geometry Aortic diameter	Optical imaging	-2.5% / <14%	Current study
Deformation circumferential strain	Optical imaging	0.0% / 5 .. 30%	Current study

The assessment of the test-retest reliability of a single measurement showed that 4D US was able to reliably distinguish between the examined load cases and deformations, even in cases where the range of the imaged deformations was very similar. Nevertheless, the results of the current study indicate, how the error of displacement and strain measurement of a single aorta or patient, can be reduced significantly: The 95% confidence interval of repeated measurements resulted in non-negligible deviations from the reference measurements. These deviations were of approximately the same size for local and average values. This indicates that the random error of repeated measurements represents the possible size of the systematic error of a single measurement. In contrast, the deviation of the median of repeated US measurements that represents the systematic error of the measurement method was of negligible size for all examined deformation metrics. Therefore, measurements of single aortic or aneurysmal wall that are intended for diagnostic use should be obtained as means of repeated examinations and evaluations of the same aortic segment. The use of strain distributions obtained from an averaged model based on ten repeated evaluations of the same US measurement could be shown to improve the reliability of the local strain estimation. Using data from averaged models, the local wall strains in calcified regions of abdominal aortic aneurysms could be distinguished from uncalcified regions significantly [24].

3D and 4D US measurement of cardiovascular geometry and deformation has been validated previously in vivo and in vitro by comparison to measurements obtained by other imaging techniques (Table 4). These studies showed a systematic deviation for the assessment of geometrical sizes similar to the results obtained in the current study, whereas the systematic deviation for measures of deformation was generally lower. The random deviation for geometry was in the range reported, as well. The random deviation of deformation measurements reported was between 6.2% and 14%, compared to values of 5% to 30% for large and small deformations, respectively.

5 Limitations of the Study

Limitations of this study are that no full 3D data were available as reference measurements and that the experimental set up did not allow for the assessment of measurement uncertainty in the x-direction that was orthogonal to the propagation direction of the RF signal and corresponded to the longitudinal direction of the examined specimen. Despite this, the experimental set up did allow the assessment of the measurement uncertainty in and orthogonal to the propagation direction (y- and z-direction, respectively). Comparably good results were obtained in both cases, indicating that the measurement uncertainty of 4D US is not direction dependent. The effect of different observers was not investigated in this study.

6 Transparency Statement

The current chapter is based on chapter 4 of the PhD thesis of Andreas Wittek [10].

References

1. Soliman, O.I.I., et al.: Accuracy and reproducibility of quantitation of left ventricular function by real-time three-dimensional echocardiography versus cardiac magnetic resonance. Am. J. Cardiol. **102**(6), 778–783 (2008)
2. Seo, Y., et al.: Validation of 3-dimensional speckle tracking imaging to quantify regional myocardial deformation. Circulation **2**(6), 451–459 (2009)
3. Park, J.-S., et al.: Validation of three-dimensional echocardiography for quantification of aortic root geometry: comparison with multi-detector computed tomography. J. Cardiovasc. Ultrasound **19**(3), 128–133 (2011)
4. Seo, Y., Ishizu, T., Enomoto, Y., Sugimori, H., Aonuma, K.: Endocardial surface area tracking for assessment of regional LV wall deformation with 3D speckle tracking imaging. JACC Cardiovasc. Imaging **4**(4), 358–365 (2011)
5. Bihari, P., et al.: Strain measurement of abdominal aortic aneurysm with real-time 3D-ultrasound speckle tracking. Eur. J. Vasc. Endovasc. Surg. **45**(4), 315–323 (2013)
6. Alessandrini, M., et al.: Detailed evaluation of five 3D speckle tracking algorithms using synthetic echocardiographic recordings. IEEE Trans. Med. Imaging (2016)

7. Derwich, W., et al.: High resolution strain analysis comparing aorta and abdominal aortic aneurysm with real time 3D speckle tracking ultrasound. Eur. J. Vasc. Endovasc. Surg. **51**(2), 187–193 (2016)
8. Karatolios, K., et al.: Method for aortic wall strain measurement with three-dimensional ultrasound speckle tracking and fitted finite element analysis. Ann. Thorac. Surg. **96**(5), 1664–1671 (2013)
9. Wittek, A., Derwich, W., Fritzen, C.-P., Schmitz-Rixen, T., Blase, C.: Towards non-invasive in vivo characterization of the pathophysiological state and mechanical wall strength of the individual human AAA wall based on 4D ultrasound measurements. ZAMM - J Appl. Math. Mech. **98**(12), 2275–2294 (2018)
10. Wittek, A.: Mechanical and Pathophysiological in Vivo Characterization of the Individual Aortic Wall Based on 4D Ultrasound Imaging. Siegen (2020)
11. Kok, A.M., et al.: On the feasibility of wall stress analysis of abdominal aortic aneurysms using three-dimensional ultrasound. J. Vasc. Surg. (2015)
12. de Hoop, H., et al.: Multiperspective Ultrasound Strain Imaging of the Abdominal Aorta. IEEE Trans. Med. Imaging **1** (2020)
13. Altman, D.G., Bland, J.M.: Measurement in medicine. Statistician **32**(3), 307 (1983)
14. Bland, J.M., Altman, D.G.: Statistical methods for assessing agreement between two methods of clinical measurement. Lancet **1**(8476), 307–310 (1986)
15. Shrout, P.E., Fleiss, J.L.: Intraclass correlations: uses in assessing rater reliability. Psychol. Bull. **86**(2), 420–428 (1979)
16. Koo, T.K., Li, M.Y.: Guideline of selecting and reporting intraclass correlation coefficients for reliability research. J. Chiropr. Med. **15**(2), 155–163 (2016)
17. Schneider, C.A., Rasband, W.S., Eliceiri, K.W.: NIH image to ImageJ: 25 years of image analysis. Nat. Methods **9**(7), 671–675 (2012)
18. Wittek, A., Fritzen, CP., Huß, A., Blase, C.: Reproducibility of in vivo constitutive parameter identification based on 4D ultrasound strain imaging. In: Tavares, J.M.R.S., Bourauel, C., Geris, L., Vander Slote, J. (eds.) Computer Methods, Imaging and Visualization in Biomechanics and Biomedical Engineering II. CMBBE 2021. Lecture Notes in Computational Vision and Biomechanics, vol. 38. Springer, Cham (2023)
19. Goss, S.A., Johnston, R.L., Dunn, F.: Compilation of empirical ultrasonic properties of mammalian tissues. II. J. Acoust. Soc. Am. **68**(1), 93–108 (1980)
20. Jones, E.M.C., Silberstein, M.N., White, S.R., Sottos, N.R.: In situ measurements of strains in composite battery electrodes during electrochemical cycling. Exp. Mech. **54**(6), 971–985 (2014)
21. Hartung, J., Elpelt, B., Klösener, K.-H.: Statistik. Oldenbourg, München (2012)
22. Li, L., Zeng, L., Lin, Z.-J., Cazzell, M., Liu, H.: Tutorial on use of intraclass correlation coefficients for assessing intertest reliability and its application in functional near-infrared spectroscopy-based brain imaging. J. Biomed. Opt. **20**(5), 50801 (2015)
23. Chen, X., et al.: 3-D correlation-based speckle tracking. Ultrason. Imaging **27**(1), 21–36 (2005)
24. Hegner, A., Wittek, A., Derwich, W., Huß, A., Gámez, A.J., Blase, C.: Using averaged models from 4D ultrasound strain imaging allows to significantly differentiate local wall strains in calcified regions of abdominal aortic aneurysms. Biomech. Model. Mechanobiol. **15**(4), 827 (2023)

Reliable Computational Fluid Dynamics for Ground Truth Generation for AI-Based Blood Flow Analysis

Benjamin F. Zwick[1(✉)], Kryspin Mirota[2], Jakub Chojnacki[2], Miłosz Gajowczyk[2], and Karol Miller[2]

[1] Intelligent Systems for Medicine Laboratory, The University of Western Australia, Perth, WA 6009, Australia
benjamin.zwick@uwa.edu.au

[2] Hemolens Diagnostics Sp. z o.o, Legnicka 48G, 54-202 Wroclaw, Poland

Abstract. In recent years the interest in computational fluid dynamics (CFD)-based analysis of blood flow in coronary arteries has intensified, largely motivated by the possibility of using reliable, comprehensive simulation results as ground truths for training AI systems potentially able to rapidly identify pathologies in hemodynamics. In this paper, we describe very detailed steady-state and transient simulations of coronary blood flow and provide recommendations on appropriate solution methods and parameters for both steady-state and transient simulations. We conclude that when using a finite volume method with polyhedral cells, the characteristic cell size of 0.075 mm is sufficient to obtain a converged, mesh-independent solution for pressure—a variable of interest in many clinically relevant applications such as estimation of the fractional flow reserve (FFR). We also recommend that for transient simulations in coronary arteries, the semi-implicit method for pressure-linked equations (SIMPLE) with 2nd order time stepping be used, with a time step of 1 ms. We suggest that for reliable time-averaged pressure drop results in coronary arteries, six cardiac cycles be simulated, with time averages taken over the last three cycles. The computational effort required for such comprehensive simulations is not prohibitive to generating results for many geometries, with the greatest bottleneck being the extraction of vessel geometries from medical images.

Keywords: Computational fluid dynamics (CFD) · blood flow · coronary arteries · AI training · hemodynamics · fractional flow reserve (FFR)

1 Introduction

Quantitative assessment of blood flow is quickly entering the clinical practice with probably the most prominent example being the assessment of the severity of coronary artery stenosis via estimation of fractional flow reserve (FFR) using computational fluid dynamics (CFD) [1–8]. FFR is a clinically validated metric that uses pressure measurements to evaluate the impact of a stenosis on blood flow, aiding in the decision-making process for interventions [6–8].

M. Kobielarz et al. (Eds.): MICCAI 2024, LNBE, pp. 16–26, 2025.
https://doi.org/10.1007/978-3-031-94128-3_3

As the main input to these simulations is the vessel geometry extracted from medical images, usually computed tomography angiography (CTA), the approach can be considered image-based, or in other words, CFD is used to extract more meaningful information from the image data.

It is therefore not surprising that to accomplish this task many research groups have turned their attention to neural network-based computer artificial intelligence (AI) systems as it is widely known that AI-based methodologies are often found very effective in generalised image analysis and information extraction from images [9, 10].

For these AI-based systems to operate reliably and accurately large amounts of training data must be provided. In the context of blood flow within the human body, it is impractical to provide such training data via comprehensive flow measurements. An obvious alternative is to conduct reliable, comprehensive computer simulations that could be considered as "full field experiments". In this paper, we describe such simulations in the context of blood flow in stenosed coronary arteries.

The remainder of this paper is organised as follows. In Sect. 2, we describe the steady-state simulation including a detailed mesh convergence study. In Sect. 3, we discuss transient analysis. Section 4 contains discussion and conclusions.

2 Steady-State Analysis of Blood Flow in Stenosed Coronary Arteries

2.1 Patient Data

We used an anonymised contrast-enhanced computed tomography angiography (CTA) image from a dataset of patients who gave their informed consent before their participation in the study, recruited as a part of a bioethics committee approved clinical trial (ClinicalTrials.gov ID NCT04777513). The case described here was selected as the geometry is sufficiently complicated to exemplify difficulties encountered by an analyst conducting the computations, see Fig. 1.

2.2 Segmentation Method

We used a U-Net-based architecture with custom (Hemolens proprietary) extensions and postprocessing for coronary vessel segmentation.

2.3 Simulation Strategy

To generate reliable results that could serve as ground truths for AI-system training, we chose the well-established and widely-used Siemens STAR-CCM + CFD software, which employs the finite volume method (FVM) for the spatial discretization of the governing equations (https://plm.sw.siemens.com/en-US/simcenter/fluids-thermal-simulation/star-ccm/). We used the segregated solver with the semi-implicit method for pressure-linked equations (SIMPLE) algorithm for pressure-velocity coupling to solve the steady-state Navier–Stokes equations.

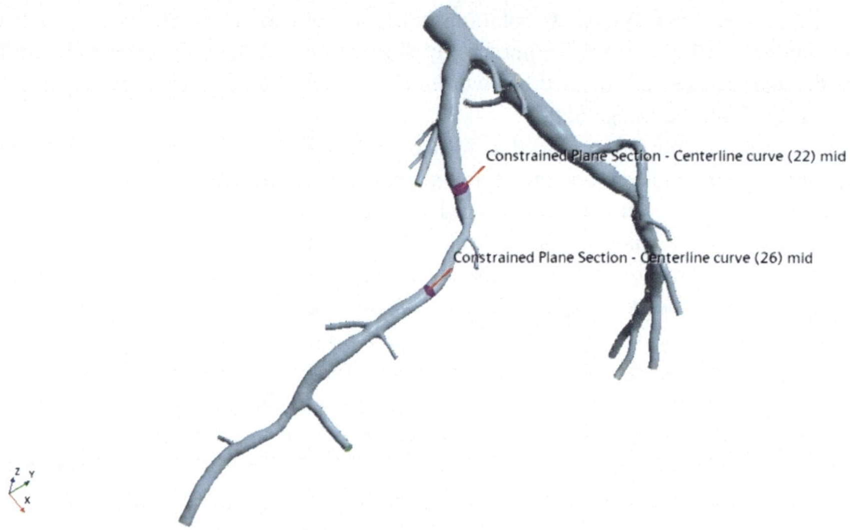

Fig. 1. Extracted geometry of analysed vessel with pressure measurement planes 22 and 26 (in front of and behind stenosis)

We ran all simulations on Setonix, an HPE Cray EX Supercomputer at the Pawsey Supercomputing Research Centre in Perth, WA, Australia (https://pawsey.org.au/systems/setonix/). We used one node with dual 2.45GHz AMD EPYC 7763 "Milan" 64-core CPU (total of 128 cores) with 256 GB RAM per simulation.

Boundary Conditions

The vessel walls were assumed to be rigid and a no-slip condition was applied at the wall boundaries.

A stagnation inlet boundary condition with a prescribed constant mean arterial pressure (MAP) of 100 mmHg was applied at the inlet of the left coronary artery (LCA).

Mass flow boundary conditions were applied at each outlet. The volumetric flow rates at the outlets were obtained by a splitting method [11] assuming a flow rate of 3 mL/s at the inlet, with the exponent equal to 2.38 based on Kassab's [12] value for the pig's LAD:

$$\frac{Q}{Q_{\max}} = \left(\frac{D}{D_{\max}}\right)^{\frac{4(\epsilon\prime+1)}{3\epsilon\prime-2}}, \epsilon\prime = 2.79, \quad (1)$$

The parameter ϵ' relates to the crown flow resistance and is equal to the ratio of metabolic to viscous power dissipation. It should be noted that there is no consensus about what value of the exponent should be used. Other groups use different exponents, e.g. (Carson et al., 2019) used a value of 2.66 and according to standard Murray's Law, the exponent is 3.

Constitutive Properties

The blood was assumed as a homogeneous, incompressible Newtonian fluid with a dynamic viscosity of 3.5 mPa·s, and mass density of 1060 kg/m^3.

Mesh Independent Solution

The first step is to establish the mesh density sufficient for calculating a mesh-independent solution. Classical mesh convergence analysis requiring structured meshes with shared nodes to directly compare nodal solutions (as stipulated by ASME V&V 40) is impractical for very complicated geometries as structured mesh refinement becomes a formidable problem then. Therefore, we compared solutions obtained with increasingly dense unstructured meshes. We concentrate on pressure because the pressure drops are used clinically to establish the severity of a stenosis through the well-established metric of fractional flow reserve (FFR). Therefore, we report minimum, average and maximum pressures in specified cross-sections as relevant variables of interest.

We considered four polyhedral finite volume meshes with characteristic lengths and cell numbers given in Table 1. To resolve the boundary layer, we used 2 boundary prism layers with a total thickness of 33% of the base size and a stretching factor of 1.5.

Table 1. Meshes used in mesh convergence study

	Mesh 1	Mesh 2	Mesh 3	Mesh 4
Characteristic length	0.3 mm	0.015 mm	0.075 mm	0.0375 mm
Number of cells	77,456	371,723	2,475,804	17,455,201

We report minimum, average and maximum pressure at outlets and selected measurement planes shown in Fig. 1.

Table 2 shows pressure values and relative differences between solutions obtained by consecutively denser meshes. The results indicate that excellent convergence has been achieved at cross-sections away from outlets, indicating that a mesh resolution of 0.075 mm is sufficient.

Figure 3 shows the pressure distribution on streamlines within the coronary arteries.

Table 2. Minimum, average and maximum pressures at two cross-sections indicated in Fig. 1 (Planes 22 and 26) and the outlets of the left anterior descending (LAD) and left circumflex (LCx) arteries for meshes with characteristic lengths of 0.3 mm; 0.15 mm; 0.075 mm; 0.0375 mm. The three coloured columns on the right show relative differences between results obtained with consecutive meshes.

Location	Pressure (mmHg)				Relative difference		
Characteristic length	0.3	0.15	0.075	0.0375	0.3	0.15	0.075
Plane 22 Min. Pres.	99.01	99.04	99.08	99.10	0.035%	0.040%	0.016%
Plane 22 Avg. Pres.	99.08	99.11	99.14	99.15	0.027%	0.034%	0.013%
Plane 22 Max. Pres.	99.12	99.15	99.18	99.20	0.028%	0.037%	0.014%
Plane 26 Min. Pres.	93.81	93.96	93.59	93.63	0.161%	0.397%	0.046%
Plane 26 Avg. Pres.	93.91	94.04	93.67	93.71	0.141%	0.399%	0.046%
Plane 26 Max. Pres.	94.00	94.20	93.85	93.88	0.213%	0.373%	0.036%
LCx Outlet Min. Pres.	95.56	95.95	96.07	95.79	0.410%	0.126%	0.291%
LCx Outlet Avg. Pres.	95.75	96.05	96.17	96.24	0.311%	0.128%	0.069%
LCx Outlet Max. Pres.	95.83	96.11	96.25	96.39	0.292%	0.147%	0.142%
LAD Outlet Min. Pres.	91.28	91.78	91.51	91.40	0.543%	0.289%	0.125%
LAD Outlet Avg. Pres.	91.33	91.81	91.57	91.66	0.524%	0.261%	0.105%
LAD Outlet Max. Pres.	91.36	91.85	91.61	92.00	0.538%	0.266%	0.424%

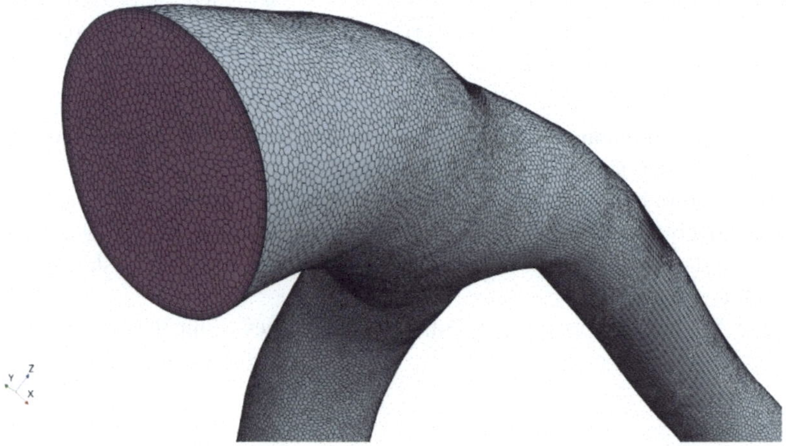

Fig. 2. Example polyhedral mesh of coronary artery bifurcation.

The results also indicate that pressure results should not be reported at outlets where the mass flow rate (or volumetric flow rate) boundary condition is imposed. Table 2 (rows 7–12) demonstrates that the convergence for pressure at the outlets is poorer than at planes located along the vessels some distance from the outlets. The iterative process attempting to reconcile the computed velocity field with the outflow boundary condition has different convergence properties than the Navier-Stokes equation solver. Therefore, pressures should be reported only at a safe distance (more than ten cell layers) from the outlet where mass or volumetric flow rate is prescribed as a boundary condition.

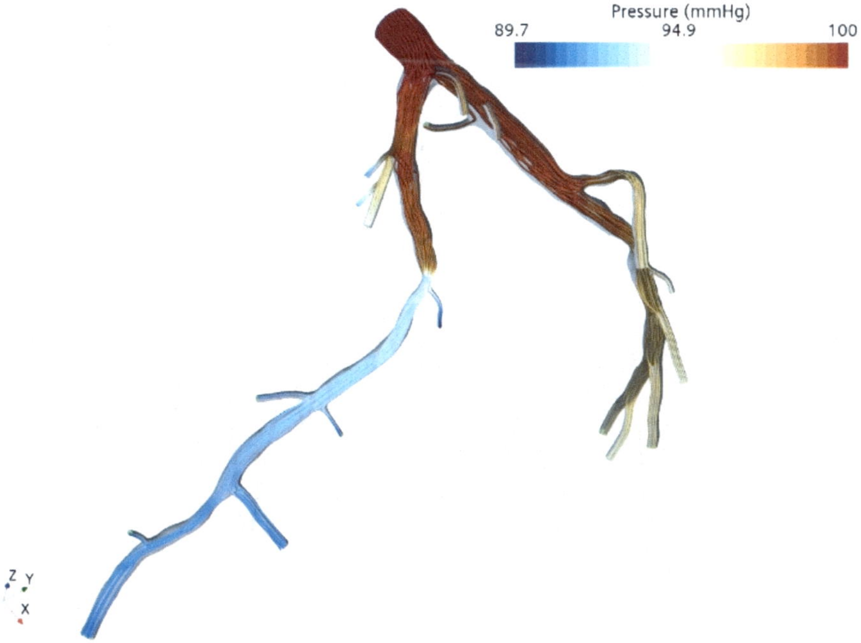

Fig. 3. Pressure distribution on streamlines for the analysed case.

3 Transient Analysis of Blood Flow in Stenosed Coronary Arteries

We conducted a comprehensive transient analysis of the blood flow in the left coronary artery shown in Fig. 1. A comprehensive transient simulation should allow capturing all scales, both spatial and temporal. The rough approximation of the required cell size is $h = L/\mathrm{Re}^{3/4}$, where L is the characteristic length (diameter) and Re is the Reynolds number. Based on measured vessel diameters and Reynolds numbers from the steady-state simulation with 3 mL/s inflow the required cell size is between 0.08 and 0.2 mm. Therefore, for transient analysis, we can safely use the same 0.075 mm resolution mesh that yields a mesh-independent solution as we demonstrated in the previous section.

3.1 Boundary Conditions

For the transient simulations, we chose boundary conditions in a similar way as for the steady-state simulations.

The vessel walls were assumed to be rigid and no-slip condition was applied at wall boundaries.

A stagnation inlet boundary condition with a prescribed arterial pressure waveform (Fig. 4) was applied at the inlet of the left main coronary artery (LCA).

Mass flow boundary conditions were applied at each outlet. The volumetric flow rates were obtained by a splitting method [11] (see Eq. 1) with a prescribed volumetric flow waveform (Fig. 4) and the exponent of 2.38 based on Kassab et al. [12].

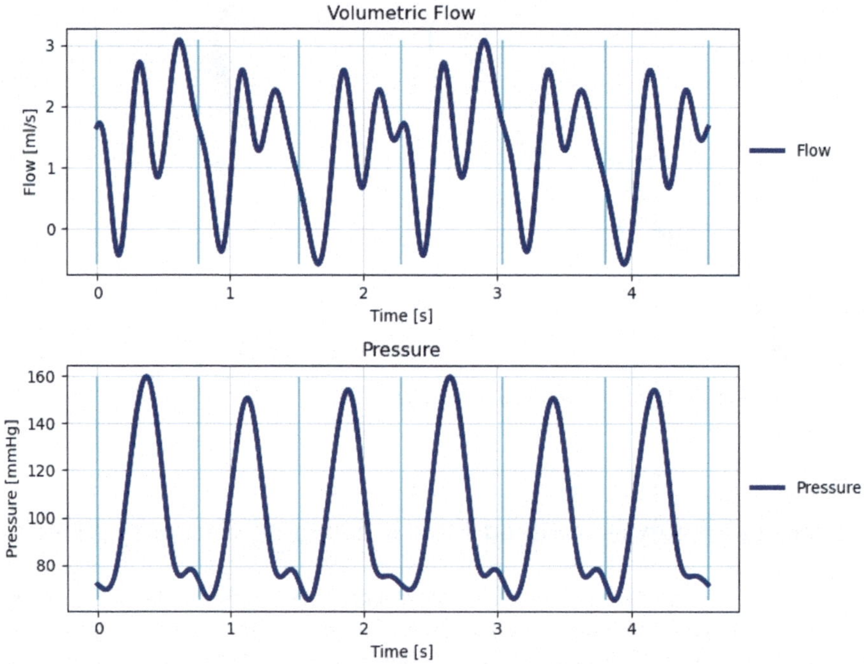

Fig. 4. Transient inflow data for the transient simulation.

3.2 Initial Conditions

We initialised flow velocities and pressure to zero.

3.3 Solution Method and Parameters

We used the semi-implicit method for pressure-linked equations (SIMPLE) with 2nd order time stepping, as recommended by the STAR-CCM + Theory Manual.

We used a time step of 1 ms, which resulted in 4750 time steps for the entire simulation (6 cardiac cycles).

We used STAR CCM + default 5 iterations per time step to converge the solution of non-linear equations in each time step. This resulted in appropriate control of residuals in each time step, see Fig. 5. It is worth noting that ascertaining convergence of the iterative process of non-linear equation solving within each time step is required by ASME V&V 20–2009 (R2021). ASME V&V 20-2009 (R2021) is in turn used by FDA Guidance "Assessing the Credibility of Computational Modeling and Simulation in Medical Device Submissions" (2023) in the procedure of credibility assessment of computational models.

We ran this simulation on a single node (128 CPU cores) of the Setonix supercomputer so that the computational effort could be directly compared to the steady-state simulations. With the parameter settings given above, the transient simulation took 2 h 15 min.

Figure 6 shows the pressure histories on the inlet and outlets and Fig. 7 displays pressures for Plane 22 in front of stenosis and Plane 26 behind stenosis (Fig. 1).

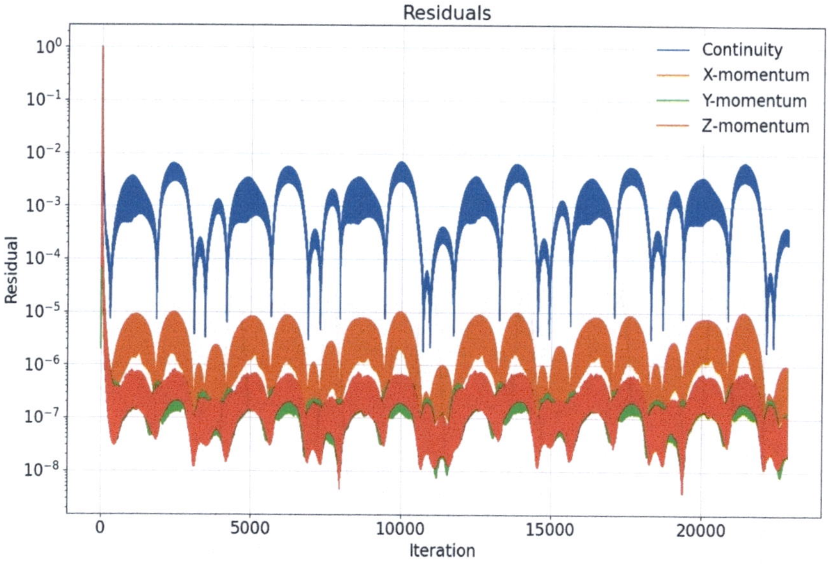

Fig. 5. Residuals over 22850 iterations (over 4570 time steps). The magnitude of residuals at the end of each time step is satisfactory.

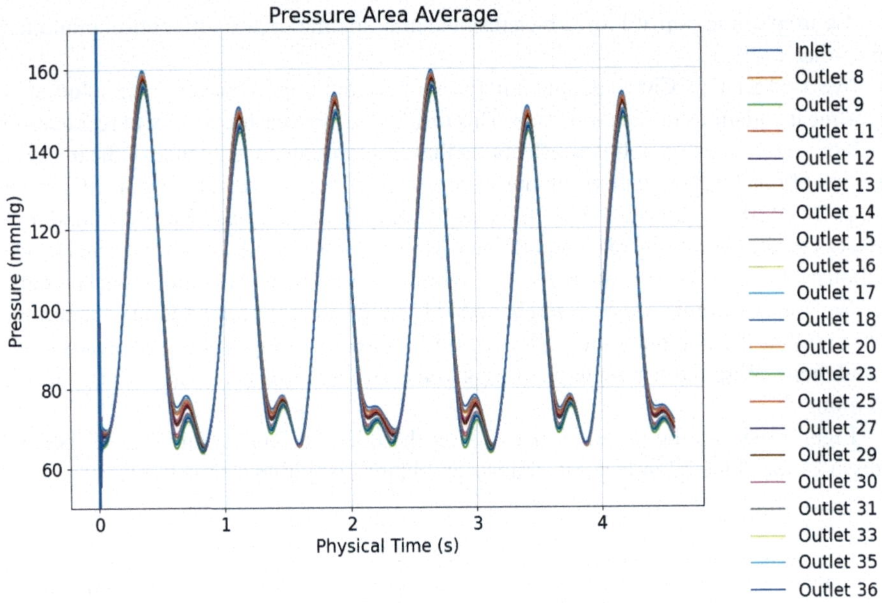

Fig. 6. Average pressure histories at inlet and outlets.

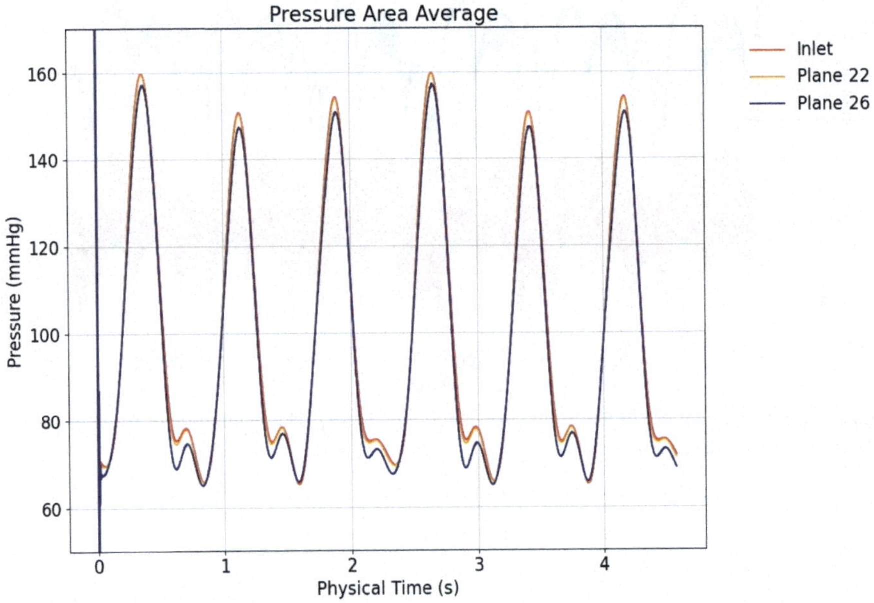

Fig. 7. Average pressure histories at Plane 22 in front of the stenosis and Plane 26 behind the stenosis.

4 Discussion and Conclusions

The presented results indicate that it is feasible to generate a large number of steady-state and transient results that can be used as bases to train AI systems. When using multiple supercomputer nodes, these simulations should run in minutes rather than hours. Furthermore, with appropriate time averaging over the last three heart cycles it is possible to generate averaged pressure drops to be used in FFR estimation.

Nevertheless, the limitations of our approach need to be highlighted. The relevance of the computed results depends on the appropriate choice of inflow and outflow conditions which is still an unresolved problem. Also, the transient results may be affected by the assumption of rigidity of arterial walls which does not depict reality accurately. Additionally, the results may depend on the specific CFD algorithm used to solve the Navier-Stokes equations. In the future, extensive parametric studies are required to resolve these inadequacies.

Acknowledgments. The authors wish to acknowledge funding from the National Health and Medical Research Council (Australia) through Ideas Grant no. APP2001689. This work was supported by resources provided by the Pawsey Supercomputing Research Centre's Setonix Supercomputer (https://doi.org/https://doi.org/10.48569/18sb-8s43), with funding from the Australian Government and the Government of Western Australia.

References

1. Sankaran, S., et al.: Uncertainty quantification in coronary blood flow simulations: impact of geometry, boundary conditions and blood viscosity. J. Biomech. **49**(12), 2540–2547 (2016)
2. Papamanolis, L., et al.: Myocardial perfusion simulation for coronary artery disease: a coupled patient-specific multiscale model. Ann. Biomed. Eng. **49**(5), 1432–1447 (2021)
3. Mittal, N., et al.: Analysis of blood flow in the entire coronary arterial tree. Am. J. Physiol.-Heart Circ. Physiol. **289**(1), H439–H446 (2005)
4. Kosior, A., Mirota, K., Tarnawski, W.: Patient - specific modeling of hemodynamic parameters in coronary arteries, USA (2018)
5. De Santis, G., et al.: Patient-specific computational fluid dynamics: structured mesh generation from coronary angiography. Med. Biol. Eng. Compu. **48**(4), 371–380 (2010)
6. Carson, J.M., et al.: Non-invasive coronary CT angiography-derived fractional flow reserve: a benchmark study comparing the diagnostic performance of four different computational methodologies. Int. J. Numer. Methods Biomed. Eng. **35**(10), e3235 (2019)
7. Morris, P.D., et al.: "Virtual" (Computed) fractional flow reserve: current challenges and limitations. JACC Cardiovas. Int. **8**(8), pp. 1009–1017 (2015)
8. Haley, H.A., et al.: Virtual (Computed) fractional flow reserve: future role in acute coronary syndromes. Front. Cardiovas. Med. **8** (2021)
9. Rygiel, P., Płuszka, P., Zięba, M., Konopczyński, T.: CenterlinePointNet++: a new point cloud based architecture for coronary artery pressure drop and vFFR estimation. In: Greenspan, H., et al. Medical Image Computing and Computer Assisted Intervention – MICCAI 2023. MICCAI 2023. Lecture Notes in Computer Science, vol. 14226. Springer, Cham (2023). https://doi.org/10.1007/978-3-031-43990-2_73
10. Pegolotti, L., et al.: Learning reduced-order models for cardiovascular simulations with graph neural networks. Comput. Biol. Med. **168**, 107676 (2024)

11. Chnafa, C., et al.: Better than nothing: a rational approach for minimizing the impact of outflow strategy on cerebrovascular simulations. Am. J. Neuroradiol. **39**(2), 337–343 (2018)
12. Kassab, G.S.: Scaling laws of vascular trees: of form and function. Am. J. Physiol. Heart Circ. Physiol. **290**(2), H894-903 (2006)

Evaluating a Differentiable Facial FEM Model with In-Vivo Simulated Data for Orthodontic Treatment Outcome Prediction

Daniel Dorda[1(✉)], Daniel Peter[2], Andrey Latyshev[3], Luís Azevedo[3], Niko Ben Huber[2], Baran Gözcü[1], Irena Sailer[3], Markus Gross[1], Bernhard Thomaszewski[1,2], and Barbara Solenthaler[1]

[1] ETH Zurich, Zurich, Switzerland
daniel.dorda@inf.ethz.ch
[2] Align Technology, Zurich, Switzerland
[3] University of Geneva, Geneva, Switzerland

Abstract. Computationally modelling the soft-tissue changes caused by orthodontics is challenging due to the small displacements caused by the treatment. Large precision is required for geometric modelling, and consequently during data collection. However, data gathered before and after treatment contains spurious facial changes from many sources, such as aging or weight-change. In this work, a novel data-collection method based on specialised dental obturators is used to obtain a clean dataset without confounding factors. The data quality is demonstrated by comparison to a cone-beam CT (CBCT) dataset of treatment outcomes. Using this dataset, we evaluate an FEM simulation with models of contact, jaw motion, and active-lip deformation, performing ablation studies to assess their impact on accuracy. We also show how differentiable simulation can optimize material parameters in the implemented models. Our results demonstrate the utility of in-vivo simulated data, and the evaluation of the simulator highlights the role of the jaw and lip in contact-driven facial deformation.

Keywords: facial soft-tissue simulation · orthodontics

1 Introduction

Several changes occur in the facial soft-tissue after orthodontic or reconstructive treatment [6,12,19]. Patients are very perceptive of such facial changes; this strongly impacts their overall evaluation of the treatment outcome [6,21].

Effectively predicting and communicating the potential changes to patients has many benefits: it enables informed decision-making in line with their healthcare priorities, instils confidence and security in their treatment, and sets realistic expectations, thus preventing unpleasant surprise or disappointment.

Finite element models are well-suited to predicting soft-tissue changes due to their precision in modelling complex physical effects and intricate geometry.

M. Kobielarz et al. (Eds.): MICCAI 2024, LNBE, pp. 27–38, 2025.
https://doi.org/10.1007/978-3-031-94128-3_4

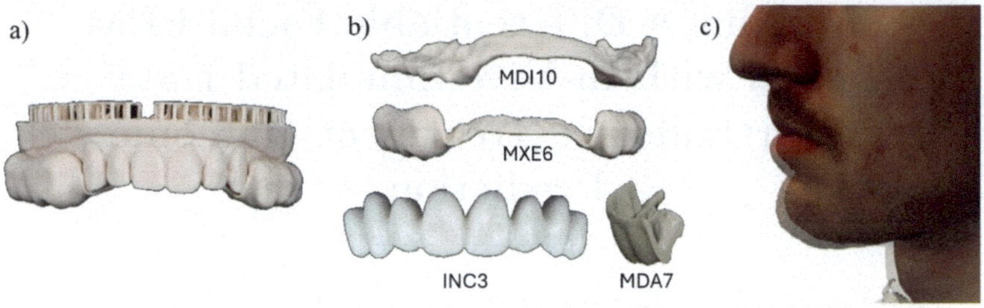

Fig. 1. (a) A photograph of the 3D printed mock-up clipping onto a tooth model. (b) Mock-up types, described further in Sect. 2. (c) An overlay of the mock-up face scan "mandibular-advancement (MDA7)" (transparent) on the neutral face scan (solid).

In graphics, generating plausible facial expressions is a core research topic. In this context, physical constraints contribute robustness and realism.

Physically-based methods have been proposed for modelling muscles [3,10], tissue pre-strains [13], lip adhesion [16] and more. Differentiable simulation is often used to optimise simulation parameters, such as muscle actuation [10] or stiffness [13], and minimise the difference between simulated and measured data.

In the medical context, facial simulation is extensively used to predict appearance after orthognathic surgery [8,14,15]. However, significantly less work has been done on predicting orthodontic outcomes [7], which involve more subtle and prolonged changes. One significant challenge in orthodontic simulation is the unreliability of paired pre- and post- treatment facial scans. The facial changes caused by the treatment are small and difficult to disentangle from those caused by growth, aging or weight-change. Such spurious, stochastic changes can easily occur during the course of treatment, which typically lasts between six months and three years. To validate a proposed simulation, a dataset of paired face scans without such confounding factors is needed, where meaningful comparisons can be made between simulation results and real-world data.

In this work, we present a dataset of "mock-up" scans - paired facial scans of individuals with and without specially designed, 3D-printed oral obturators, shown in Fig. 1. The mock-up scans closely mimic the effect of many common orthodontic treatments, while removing the aforementioned time-scale dependent issues, since the scans are captured in one continuous session [2]. Predicting the effects of mock-ups on soft-tissue becomes a surrogate problem and a prerequisite for predicting orthodontic outcomes. Moreover, it provides a clean dataset on which optimisation methods may be used. We design five types of mock-ups that imitate different types of changes.

Subsequently, we develop and test an FEM simulation of facial soft tissue on our mock-up data. Our simulation includes the contact between lips and mock-ups, a force for lip-closure, and jaw articulation. Through its differentiable formulation it allows for the optimisation of arbitrary simulation parameters. Here, we test the optimisation over heterogeneous material stiffness. To our knowledge this is the first test of differentiable simulation for orthodontics, and we highlight several key challenges with such an approach, including the problem of generalisation across individuals and treatment types.

In the remainder of our paper, we start by outlining the dataset and data-processing pipeline. Next, we explain the simulation, first detailing the design of the forward solver and its constituent models, followed by the inverse solver and its objective function. In the results section, we show our analysis of errors in mock-up and CBCT treatment data, present the forward-simulation results and comment on the performance for the various mock-up types, and finally we cover the experiments on differentiable simulation. We aim for this work to demonstrate the major challenges in orthodontic soft-tissue simulation and propose a methodology for overcoming these challenges.

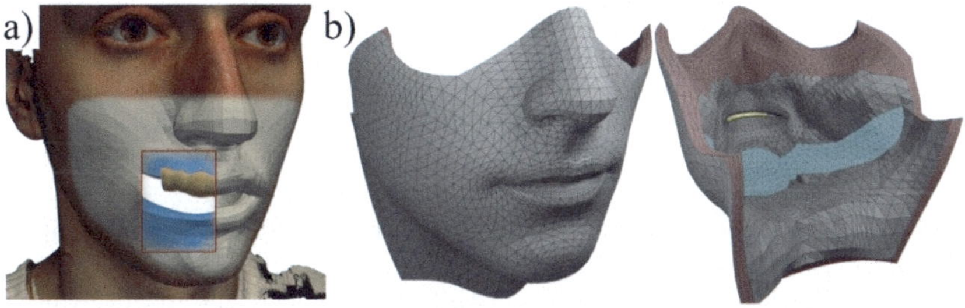

Fig. 2. (a) The intermediate stages of data-processing: the raw face mesh (textured), registered face mesh (white), registered bones (blue) and mock-up (orange). (b) A tetrahedral face mesh. Coloured regions show: zero-displacement surface (red), jaw-force application surface (blue), lip-force application surface (yellow).

2 Data

For each participant in study, we collect: six 3D face scans, five mock-ups, a CBCT scan, and an intraoral (IO) scan. In total, the dataset consists of five participants, four male and one female. To acquire the mock-up data, we use an high-resolution 3D scanner [4], a commercially available IO scanner, and a CBCT scanner with a 0.3 mm resolution for all axes. An additional dataset of 180 CBCT scan pairs is used to validate the quality of mock-up data.

Five classes of mock-up are designed in Exocad and 3D printed (Fig. 1). We designate the five types based on their effect as follows: INC3 – a 3 mm incisor advancement, INC6 – a 6 mm incisor advancement, MDI10 – a 10 mm inferior mandibular movement, MDA7 – a 7 mm anterior mandibular movement, MXE6 – a 6 mm maxillary expansion. The mock-ups contain slots for upper and lower teeth, pushing on the upper lip and forcing the mandible into a prescribed position. We capture the 3D deformed face for each mock-up, and an additional scan of the neutral expression.

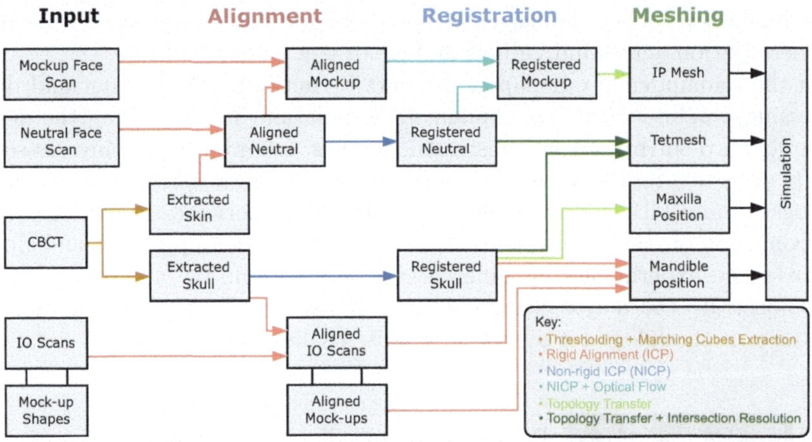

Fig. 3. Flowchart of the data-processing pipeline. Nodes represent intermediate processed states, and arrows represent the processes between them.

To process scan data into a set of simulation-ready meshes, we use a pipeline, which consists of three major steps: *alignment*, *registration*, and *meshing*. The flow chart in Fig. 3 details the process, while Fig. 2 shows the data during processing. The data produced by the pipeline are: a tetrahedral mesh of the soft-tissue, a list of nodes on the maxilla surface which become a zero-displacement BC, a list of nodes on the lips, a list of nodes close to the mandible, a rigidly transformed mandible mesh, the mock-up mesh, and mock-up face scan. Since the soft-tissue affected by treatment is limited to the lower half of the face, we simplify the simulation by cutting the mesh off under the eyes, as seen in Fig. 2. On average, our simulation meshes have 12k nodes and 60k elements.

The quality of the outputs depends on several factors. Accurate alignment and registration are crucial for obtaining meaningful correspondences between neutral and mock-up scans. We find that a texture-based registration step, which relies on small, high-frequency features like spots and blemishes, is necessary to guide the process in geometrically smooth areas of the face like the cheeks, chin, neck or nose. In total, the registration process introduces negligible errors of 0.05 ± 0.01 mm on the facial surface.

Soft-tissue thickness can also influence the simulation output. In our method, the soft-tissue conforms to the bony surface of the mandible and cranium. The tissue thickness of lips, cheeks, and other regions which aren't attached to the bone is set automatically via the registration process [1]. Some meshes are manually post-processed such that the lips are in contact with the teeth. This approach is fast and adapts well to new identities, but the precise control over tissue thickness can be improved.

The CBCT-only data is processed in a similar manner, using the extracted skull for rigid alignment between pre- and post- treatment pairs, and thresholding followed by NICP [1] to obtain registered surfaces.

3 Simulation

We simulate the deformation of the facial soft-tissue caused by the contact between the mock-up and the lip, the accompanying movement of the jaw, and a force that mimics active closure of the lower lip, a reflex which occurs under certain mock-ups. The problem is formulated as an energy minimisation using the finite element method.

The total energy of the system, $E_T(\mathbf{x}, \mathbf{X}, \alpha)$, depends on the deformed- and rest-state node positions, and the position of the mock-up respectively. This energy is a sum of four terms:

$$E_T = E_{NH}(\mathbf{x}, \mathbf{X}) + E_C(\mathbf{x}, \alpha) + E_J(\mathbf{x}) + E_L(\mathbf{x}) \quad . \tag{1}$$

These are: the internal Neo-Hookean elastic deformation energy E_{NH} [20], the contact barrier energy between the mock-ups and the lip E_C, and two spring-like energies for jaw articulation E_J and lip-position restoration E_L. We use a tissue density of 0.9 g/ml, and a Poisson ratio of 0.47. For Sect. 4.2 we set the Young's modulus to 10kPa, before optimising it freely in Sect. 4.3.

We find the quasistatic, minimum energy state where $\frac{\partial E_T}{\partial \mathbf{x}} = \mathbf{0}$ using Newton's method with line search. The Jacobian and Hessian of E_T are calculated analytically, and derived via symbolic differentiation software [18]. We use a Cholesky solver from the Eigen library [9] and apply diagonal regularisation when encountering an indefinite Hessian matrix.

Contact Energy. We implement an intersection-free model of contact based on the barrier-energy formulation described by Li et al. [17]. Under this model, a node $\mathbf{x}_i$ of $\mathbf{x}$, at a distance $d_i(\mathbf{x}_i, \alpha)$ to the mock-up accrues an energy of

$$E_{Ci} = -(d_i - \hat{d})^2 \ln(\frac{d_i}{\hat{d}}) \tag{2}$$

if $d_i(\mathbf{x}_i, \alpha) < \hat{d}$, and 0 otherwise. We calculate $d_i(\mathbf{x}_i, \alpha)$ using an IMLS field [23], which is differentiable with respect to $\mathbf{x}$. We set the threshold $\hat{d} = 0.02$.

The parameter α controls the size and position of the mock-up. At $\alpha = 1$, the mock-up is in the correct position, and at $\alpha = 0$ it is in a fictitious, intersection-free initial position. We incrementally increase α to 1. This procedure is necessary, since the simulation must be initialised in an intersection free state.

To reduce the number of queries to the IMLS field we implement a broad-phase near-collision detection using an SDF [11], which filters all nodes with distances $> 1.1\hat{d}$.

Deformation Energy. We model the effects of mandibular movement and lip-closure via a spring-like energy. For a subset of the nodes in $\mathbf{x}$, extracted via a selector matrix S, we assign a target position $\hat{\mathbf{x}}_{\mathrm{T}}$, and apply a quadratic penalty

$$E = \frac{1}{2} k \|\mathrm{S}\mathbf{x} - \hat{\mathbf{x}}_T\|_2^2 \ . \tag{3}$$

This model allows us to apply"soft-constraints", whose strength can be adjusted via the stiffness k. For the jaw model, we set a high-value of k, which deforms the mandibular nodes to within 0.15mm of the target position while removing the need for complex iterative loading schemes. For the lip model, we use a spatially-varying value of k, with $k = 1.5$ closest to the centre, and $k = 0.75$ for nodes on the boundary of the lip region. This allows the deformation in the lip region to be smoother and more realistic.

The target positions for the jaw are specified from a rigid transform which aligns the lower teeth into the proper position with respect to the mock-up type. For the lips, the target-position of the lower lip is set as their initial closed position, while the upper lip is allowed to deform freely.

3.1 Inverse Solver

We extend our solver with an inverse mode, which uses sensitivity analysis and the simulation gradients to minimise an objective function $T(\mathbf{x}, \mathbf{p}, \hat{\mathbf{x}}_{\text{gt}})$ over an arbitrary parameter set, where $\hat{\mathbf{x}}_{\text{gt}}$ is the ground truth face shape (with mock-up), and $\mathbf{p}$ is a vector of optimisation parameters. In this work, we minimise the simulation error, i.e. the difference between output and ground truth, by optimising over per-element Young's modulus. Thus $\mathbf{p} \in \mathbb{R}^{|\mathcal{E}|}$, where $\mathcal{E}$ denotes the set of tetrahedral elements. We define the objective function as

$$T(\mathbf{x}, \mathbf{p}, \hat{\mathbf{x}}_{\text{gt}}) = T_x(\mathbf{x}, \hat{\mathbf{x}}_{\text{gt}}) + \lambda_1 T_{lap}(\mathbf{p}) + \lambda_2 T_{bar}(\mathbf{p}) \tag{4}$$

where $T_x(\mathbf{x}, \hat{\mathbf{x}}_{\text{gt}})$ is a quadratic penalty, like the one defined in Eq. 3, and the terms T_{lap} and T_{bar} are Laplacian and barrier regularisation. Regularisation is necessary to make the problem well-posed and to guarantee stability.

We use an SGN solver [22] to accelerate the optimisation. Since all terms in the simulation energy and objective function are C^2 smooth w.r.t. $\mathbf{x}$ and $\mathbf{p}$, we can easily calculate the analytical gradients required for the solver as before.

Laplacian Regularisation. The Laplacian matrix $\mathrm{L} \in \mathbb{R}^{|\mathcal{E}|\times|\mathcal{E}|}$ captures the local connectivity of the tetrahedral elements, representing how each element is connected to its neighbours. The entries of the Laplacian are defined such that the diagonal entries represent the degree (number of neighbours of each tetrahedron), and the off-diagonal entries are negative weights representing the connections between elements. We adapt and vectorise the Laplacian regularisation of Bickel et al. [5], and calculate the penalty as:

$$L_{\text{Lap}} = \frac{\lambda}{2} \|\mathrm{L}\mathbf{p}\|_2^2 \ . \tag{5}$$

We set $\lambda = 10^{-9}$, which pushes the optimiser towards smooth solutions. High values of λ cause premature convergence, and low values lead to instability.

Barrier Regularisation. The barrier regularisation prevents the optimiser from converging to out-of-bounds solutions, i.e. ones with negative stiffness. The regularisation function is defined as

$$L_{\text{Barr}} = \sum_{e \in \mathcal{E}} -(p_e - \hat{p}_u - \hat{p}_l)^2 \ln(\frac{p_e - \hat{p}_u}{\hat{p}_l}) \text{ , iff } p_e < \hat{p}_u + \hat{p}_l \text{ else } 0 \text{ ,} \tag{6}$$

where p_e is the parameter value of element e, $\hat{p}_u$ is the lowest allowed parameter value and $\hat{p}_l$ is a threshold at which regularisation becomes non-zero. We set these to 100Pa and 500Pa respectively.

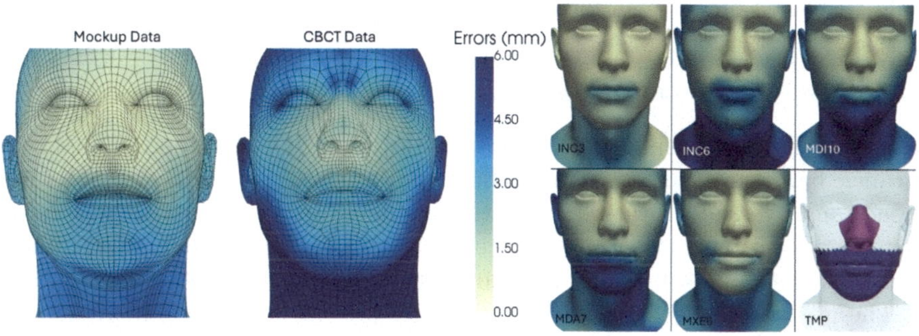

Fig. 4. Left: Comparison of average facial changes in mock-up data and CBCT pre-post treatment data. Right: Average changes of each mock-up. The TMP represents the facial areas used to calculate errors on the face (purple) and nose (pink).

4 Experiments

We first analyse our dataset and compare it against a CBCT dataset of before-after treatment scans. Then, we run forward simulations with the mock-ups. We compare the performance across the mock-up types, and show the importance of the jaw and lip effects via an ablation study. Finally, we show the impact of material parameters by evaluating the performance of our inverse simulation.

4.1 Dataset Analysis

To quantify the deformation applied by the mock-ups, we calculate per-vertex distance between the neutral and deformed face scans and average it over all subjects. For the pre-post CBCT data, we similarly calculate the average treatment effect. As seen in Fig. 4, the mock-up data shows a lower average displacement of 2.50 ± 1.60 over the simulation region, compared to 3.74 ± 3.10 in the CBCT. However, measuring the error only on the nose region, we obtain 0.77 ± 0.44 mm for mock-ups and 1.90 ± 1.28 mm for CBCT.

While still non-zero, the lower error indicates the mock-up data contains better correspondences, since in the nose region, which does not change due to neither mock-up nor treatment, the error can only be explained by misalignment and vertex-drift. Vertex-drift is a large source of inaccuracy in paired CBCT scans, where dense landmarking of smooth, featureless skin surfaces such as the nose, cheeks and neck is very error prone. In mock-up data, facial key-points can be tracked with high-precision on the high-resolution skin textures. A subsequent optically-based mesh re-registration then reduces the amount of drift between the scans. In real pre-post treatment data, where scans are taken months apart, such correspondences cannot be relied upon. To improve the quality of the mock-up data further, a jaw-tracking method could be used during scanning, and markers could be placed on the skin for simplified global alignment.

4.2 Forward Simulation

To evaluate the performance of the simulator, we calculate the mean Euclidean distance between corresponding vertices of the simulation and ground truth meshes and average it over all subjects. Examples of the simulation results are shown in Fig. 5, and the error measurements are contained in Table 1. These simulations use uniform material parameters as defined in Sect. 3.

The ablation study reveals that removing jaw and lip effects increased the average error from 2.1±0.6 mm to 2.5±1.5 mm and 2.4±1.5 mm respectively. This highlights the crucial role that the lip and jaw play in determining face shape - and suggests that more research is needed to develop precise modelling capability of lip response in orthodontic treatment.

In terms of absolute error, the worst performance is obtained by the INC6 mock-up. Compared to other mock-ups like MDI10 and MDA7, we observed that the jaw position for INC6 was more varying between subjects, and due to the large deformation, several subjects tried to actively close their lips by flexing the muscles of their chin - an effect only partially accounted for by the lip-force. Additionally, mock-ups with the lowest initial change (INC3 and MXE6) were least improved by the simulation. This is likely explained by these more subtle mock-ups being more sensitive to imprecise modelling of intraoral geometry - an area that requires further study.

Table 1. Mean millimetres of error for each mock-up type.

	INC3	INC6	MDI10	MDA7	MXE6	Mean
No Simulation	1.9 ± 0.6	3.3 ± 0.3	3.4 ± 0.9	2.4 ± 0.9	1.7 ± 0.5	2.5 ± 1.6
Full Simulation	2.0 ± 0.9	2.9 ± 1.5	1.9 ± 1.1	2.0 ± 1.4	1.7 ± 0.9	2.1 ± 0.6
No Lip Force	2.4 ± 1.4	3.3 ± 1.9	2.4 ± 1.4	2.2 ± 1.4	1.9 ± 1.0	2.4 ± 1.5
No Jaw Force	1.9 ± 0.9	2.9 ± 1.2	3.4 ± 2.0	2.2 ± 1.4	1.9 ± 1.1	2.5 ± 1.5

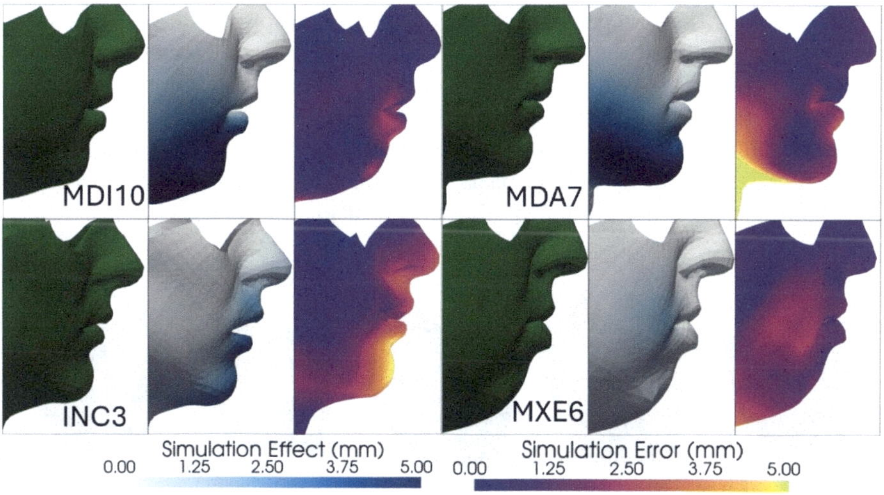

Fig. 5. Examples of four simulations, showing the initial simulation mesh (left), the simulated outcome colour-coded by displacement from the rest-shape (middle), and the ground-truth mesh colour-coded by the error (right).

4.3 Inverse Simulation

We optimise heterogeneous material stiffness (Young's Modulus), with the aim of measuring whether parameter values optimised for different individuals and different mock-ups share similarities, and could potentially be applied to unseen cases. Our experiments show that the optimised parameters slightly increase accuracy for the specific mock-up simulation they are fitted to, however we do not observe strong generalisation between individuals or different mock-up types.

Concretely, we jointly optimise a stiffness that minimises the objective for mock-up MDA10 and INC6. We also optimise individually for minimising loss on MDI10 only. Figure 6 shows the material parameter distributions. On average, the joint optimisation decreased error by 7.4%, while single-mock-up optimisation decreased the error by 19.6%. When optimised parameters were tested against mock-ups not used during optimisation, namely INC3, MDA7 and MXE6, the average error decreased by 1.7%.

These findings suggest that the major sources of simulation errors are different for each case and individual, and the optimisation finds an over-fitted solution for each problem. This explanation also supports the finding that improvement was lower when one parameter set was optimised for multiple mock-ups at once. Additionally, the distribution of the parameters is not constrained by and does not resemble biological tissue properties. This further supports the argument that the optimisation process attempts to over-correct a mixture of errors.

The limitations of this simulator – lack of muscle-driven lip actuation, unmeasured intra-oral soft-tissue geometry, or isotropic tissue model – are all a potential source of error. More precise, optimised simulation models could attain lower errors. However, as more parameters are considered for optimisation, their interactions can complicate the problem. For example, apparent tissue stiffness may be an effect of both Young's modulus and tissue thickness. Jointly optimising both parameter vectors may therefore require advanced approaches to guarantee good solutions. Achieving generalised accuracy may also require for parameters to be generated on an individual basis, for which machine-learning models may prove suitable. We leave these experiments to future work.

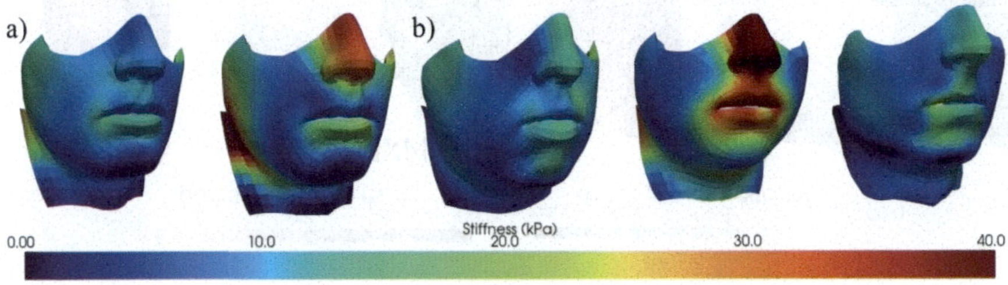

Fig. 6. Optimised parameter maps. (a) The same subject, optimised jointly for MDI10 and INC6 (left), and optimised only for MDI10 (right). (b) 3 individuals jointly optimised for MDI10 and INC6. Differences in the optimised maps may come from the diverse sources of error specific to each simulation.

5 Conclusion

Delivering reliable outcome predictions is essential for enhancing patient experience. To develop accurate orthodontic simulations, it is crucial to obtain high-quality datasets and create biomechanically valid yet flexible models, capable of capturing the individual variations in facial mechanics.

The deformations in our mock-up dataset represent a simplified subset of real treatment changes. Solving the mock-up simulation problem is a necessary step towards fully realised orthodontic simulations. The more precise point correlations provided by the mock-up data enable a meaningful, systematic evaluation of the simulation model, and areas for refinement can be more easily identified.

With reliable data, future research can focus on testing advanced biomechanical models and optimisation tools for parameter discovery. Important avenues include muscle models for active lip closure or temporomandibular joint constraints. Additionally, optimization methods can explore the complex parameters spaces of such models, discovering critical parameters, and methods for measuring or inferring these parameters for unseen individuals may be researched.

With ongoing innovation, orthodontic soft-tissue simulation can become a staple of dental treatment. The benefits it confers to practitioners and patients provide a strong motivation for continued advancement in this challenging area.

Acknowledgements. This research was funded by Align Technology Switzerland GmbH (project AlignTech-ETH).

References

1. Amberg, B., Romdhani, S., Vetter, T.: Optimal step nonrigid ICP algorithms for surface registration. In: 2007 IEEE Conference on Computer Vision and Pattern Recognition, pp 1–8 (2007). https://doi.org/10.1109/CVPR.2007.383165
2. Au, J., Mei, L., Bennani, F., Kang, A., Farella, M.: Three-dimensional analysis of lip changes in response to simulated maxillary incisor advancement. Angle Orthod. **90**(1), 118–124 (2020)
3. Bao, M., Cong, M., Grabli, S., Fedkiw, R.: High-quality face capture using anatomical muscles. In: IEEE Conference on Computer Vision and Pattern Recognition, CVPR 2019, Long Beach, CA, USA, June 16-20, 2019, Computer Vision Foundation, pp. 10802–10811. IEEE (2019). https://doi.org/10.1109/CVPR.2019.01106
4. Beeler, T., Bickel, B., Beardsley, P., Sumner, B., Gross, M.: High-quality single-shot capture of facial geometry. ACM Trans. Graph. Proc. SIGGRAPH **29**(3), 40:1-40:9 (2010)
5. Bickel, B., Bächer, M., Otaduy, M.A., Matusik, W., Pfister, H., Gross, M.: Capture and modeling of non-linear heterogeneous soft tissue. ACM Trans. Graph. **28**(3) (2009). https://doi.org/10.1145/1531326.1531395
6. Bravo, L.A., Canut, J.A., Pascual, A., Bravo, B.: Comparison of the changes in facial profile after orthodontic treatment, with and without extractions. Br. J. Orthod. **24**(1), 25–34 (1997)
7. Chen, S., Lou, H., Guo, L., Rong, Q., Liu, Y., Xu, T.M.: 3D finite element modelling of facial soft tissue and preliminary application in orthodontics. Comput. Methods Biomech. Biomed. Engin. **15**(3), 255–261 (2012)
8. Dorda, D., et al.: Differentiable simulation for outcome-driven orthognathic surgery planning. Comput. Graph. Forum **41**(8), 53–61 (2022). https://doi.org/10.1111/cgf.14623
9. Guennebaud, G., Jacob, B., et al.: Eigen v3 (2010). http://eigen.tuxfamily.org
10. Ichim, A.E., Kadlecek, P., Kavan, L., Pauly, M.: Phace: physics-based face modeling and animation. ACM Trans. Graph. **36**(4), 153:1-153:14 (2017). https://doi.org/10.1145/3072959.3073664
11. Jacobson, A., Kavan, L., Sorkine-Hornung, O.: Robust inside-outside segmentation using generalized winding numbers. ACM Trans. Graph. Proc. ACM SIGGRAPH **32**(4), 33:1-33:12 (2013)
12. Janson, G., Mendes, L.M., Junqueira, C.H., Garib, D.G.: Soft-tissue changes in Class II malocclusion patients treated with extractions: a systematic review. Eur. J. Orthod. **38**(6), 631–637 (2016)
13. Kadleček, P., Kavan, L.: Building accurate physics-based face models from data. Proc. ACM Comput. Graph. Interact. Tech. **2**(2) (2019). https://doi.org/10.1145/3340256
14. Kim, D., et al.: A new approach of predicting facial changes following orthognathic surgery using realistic lip sliding effect. MICCAI **11768**, 336–344 (2019)
15. Koch, R.M., Gross, M.H., Carls, F.R., von Büren, D.F., Fankhauser, G., Parish, Y.I.H.: Simulating facial surgery using finite element models. In: Proceedings of the 23rd Annual Conference on Computer Graphics and Interactive Techniques, Association for Computing Machinery, New York, NY, USA, SIGGRAPH '96, pp. 421–428 (1996). https://doi.org/10.1145/237170.237281

16. Leach, M., Maddock, S.: An evaluation approach for a physically-based sticky lip model †. Computers **8**(1) (2019). https://doi.org/10.3390/computers8010024
17. Li, M., et al.: Incremental potential contact: intersection- and inversion-free large deformation dynamics. ACM Trans. Graph. SIGGRAPH **39**(4) (2020)
18. Maple: Maplesoft, a division of Waterloo Maple Inc. (2019). https://www.maplesoft.com/
19. Pan, Y., Chen, S., Shen, L., Pei, Y., Zhang, Y., Xu, T.: Thickness change of masseter muscles and the surrounding soft tissues in female patients during orthodontic treatment: a retrospective study. BMC Oral Health **20**(1), 181 (2020)
20. Sifakis, E., Barbic, J.: Fem simulation of 3D deformable solids: a practitioner's guide to theory, discretization and model reduction. In: ACM SIGGRAPH 2012 Courses, Association for Computing Machinery, New York, NY, USA, SIGGRAPH '12 (2012). https://doi.org/10.1145/2343483.2343501
21. Zarif Najafi, H., Sabouri, S.A., Ebrahimi, E., Torkan, S.: Esthetic evaluation of lip position in silhouette with respect to profile divergence. Am. J. Orthod. Dentofacial Orthop. **149**(6), 863–870 (2016)
22. Zehnder, J., Coros, S., Thomaszewski, B.: SGN: sparse Gauss-Newton for accelerated sensitivity analysis. ACM Trans. Graph. **41**(1), 1–10 (2022). https://doi.org/10.1145/3470005
23. Öztireli, A.C., Guennebaud, G., Gross, M.: Feature preserving point set surfaces based on non-linear kernel regression. Comput. Graph. Forum **28**(2), 493–501 (2009). https://doi.org/10.1111/j.1467-8659.2009.01388.x

The Comparison of 2D and 3D Based Models for the Problem of Plaque Segmentation and Coronary Artery Calcium Scoring on Non-contrast Cardiac CT Imaging

Jakub Chojnacki[1,2(✉)], Miłosz Gajowczyk[1], Kinga Teklak[1], Tomasz Konopczyński[1], and Maciej Zamorski[1]

[1] Hemolens Diagnostics Sp. z o.o, Legnicka 48G, 54-202 Wroclaw, Poland
jakub.chojnacki@hemolens.com

[2] Wroclaw University of Science and Technology, Wyb. Wyspiańskiego 27, 50-370 Wrocław, Poland

Abstract. Calcium scoring is a fundamental technique for assessing the extent of calcification in the cardiovascular system, which can be used for diagnostic purposes and predicting the occurrence of major adverse cardiac events (MACE). Although it is a standard procedure, the process is quite challenging and ambiguous, often resulting in varying results between different specialists. In addition, manual identification of coronary calcification is time-consuming and impractical on a large scale, thus increasing the demand for automated methods. This work aims to evaluate various deep learning models and forms of data processing (segmenting per 2 d slice vs using 3 d sliding window) used for the problem of fully-automatic plaque segmentation from non-contrast cardiac CT images with subsequent plaque volume and Agatson score estimation.

Keywords: calcium scoring · Agatston score · computed tomography · deep learning · plaque detection

1 Introduction

Automatic Coronary Artery Calcium Scoring (CACS) methods that use cardiac computed tomography (CT) are vital for the diagnosis and treatment of heart disease. These methods aim to identify and quantify coronary artery calcium (CAC) lesions, providing crucial information about atherosclerotic plaques in patients without symptoms. The Agatston scale [2], commonly used in this context, quantifies arterial calcification, with higher scores indicating an increased risk of cardiovascular disease [6]. Despite its clinical value, the Agatston scale has limitations, particularly in detecting low-density non-calcified ("soft") atherosclerotic plaques lacking calcium [4]. A standard approach to the analysis of coronary vessel calcification involves setting a threshold to identify lesions with densities of 130 Hounsfield units (HU) or greater in areas of at least 1 mm^2 [2]. These criteria help mitigate noise and reduce the impact of individual pixel variations, ensuring that only significant calcifications are included in the analysis.

M. Kobielarz et al. (Eds.): MICCAI 2024, LNBE, pp. 39–49, 2025.
https://doi.org/10.1007/978-3-031-94128-3_5

Performing a precise plaque scoring is a challenging task, primarily due to the intricate nature of the spatial reconstruction performed by CT scanners [7, 23], particularly for imagining in the cardiac region. In order to reduce patient exposure to radiation, images captured for calcium scoring purposes are made with low spatial resolution (usually with 3 mm slice thickness) and minimal radiation dose (0.8mSv to 10.5mSv [11]). Although safer for the patient, this imaging protocol can produce motion artifacts and distort the size of small objects due to the fairly large size of the voxel, resulting in an imprecise final score.

Another major flaw is an inability to perform calcium scoring automatically. Most currently available software packages for semi-automatic CS perform thresholding on 130 HU regions and require the user to manually select plaque areas in order to compute the final Agatston score. For low-resolution, low-dose images, this requires attention of the radiologist and highlights the need for fully-automatic solutions for calcium estimation.

To address these challenges, we compare several popular machine learning approaches to semantic segmentation with application to automatic Agatston score and plaque volume approximation task. We present our findings for both 2 D and 3 D image types.

2 Related Work

The application of artificial intelligence (AI) in cardiovascular imaging has gained significant attention in recent years, particularly in the evaluation of coronary artery calcium (CAC) [12, 13]. CAC scoring techniques are typically categorized into semi-automatic and fully automated methods.

Semi-automatic techniques combine manual identification and quantification of lesions with software assistance, enhancing the precision and reliability of evaluations. For example, methods that integrate manual and automated processes have been proposed to improve accuracy and efficiency in CAC detection [24, 26]. These approaches benefit from human expertise while leveraging computational power to streamline the scoring process.

Fully automated techniques, on the other hand, employ machine learning algorithms to provide efficient processing and reliable prognostic information on coronary artery disease [14]. These methods are essential for clinical applications because of their ability to handle large datasets and deliver consistent results. Automated CAC scoring methods are evaluated by comparing them with expert manual scoring or by analyzing test datasets with known calcification patterns. Common evaluation metrics include sensitivity, specificity, accuracy, and the area under the receiver operating characteristic curve (ROC AUC) [16, 27, 10].

Validation methods such as leave-one-out cross-validation [3] are crucial to ensuring the robustness of these algorithms across various datasets. These methods help to assess both general precision and local precision in detecting calcium deposits in the coronary arteries. Such rigorous evaluations help determine the applicability of automated CAC scoring methods in diverse demographics of patients and severities of the disease, highlighting their potential for broader clinical implementation [21,10].

Furthermore, evaluation frameworks often include evaluations per patient and per artery, which are essential to understand the detailed performance of automated systems. These evaluations involve comparing automated scores with manual benchmarks to measure the precision of quantifying calcium deposits [1]. Using these comprehensive validation strategies, researchers can ensure that automated CAC scoring methods are reliable and effective for clinical use.

3 Proposed Approach

Accurately segmenting plaques on CT images is challenging due to the similar intensity between calcium-based plaques and bone structures, and this issue is further complicated by artifacts. To resolve these issues, a systematic approach was introduced. Initially, a thresholding method was used to isolate relevant structures with Hounsfield units (HU) above 130:

$$I'(x, y, z) = \begin{cases} I(x, y, z) & \text{if } I(x, y, z) > 130 \\ 0 & \text{otherwise} \end{cases} \tag{1}$$

For better contextual understanding, anatomical regions such as the aorta, pulmonary artery, and heart chambers were segmented using the TotalSegmentator [25] and added to the ground truth.

Additionally, aortic plaques were excluded from further processing:

$$\text{Mask}_{\text{aorta}}(x, y, z) = \begin{cases} 0 & \text{if voxel is part of aortic plaque} \\ I(x, y, z) & \text{otherwise} \end{cases} \tag{2}$$

The segmentation pipeline utilized multiple channels. The plaques were segmented in the first channel, the aortas in the second, and other anatomical structures in the last channel.

$$f(x, y, z) = \begin{cases} 0 & \text{if pixel is part of background} \\ 1 & \text{if pixel is part of plaques} \\ 2 & \text{if pixel is part of aorta} \\ 3 & \text{if pixel is part of other mentioned structures} \end{cases} \tag{3}$$

To reduce checkerboard artifacts in high-resolution images, a method combining nearest-neighbor interpolation followed by convolution was applied:

$$I_{\text{upsampled}}\left(x', y', z'\right) = \sum_{x,y,z} I(x, y, z) \cdot K\left(x' - x\right) \cdot K\left(y' - y\right) \cdot K\left(z' - z\right) \tag{4}$$

where $I_{\text{upsampled}}(x', y', z')$ represents the interpolated intensity and K is the interpolation kernel.

To tackle the spatial resolution problems of CT scanners, various methods incorporating distinct input data formats (either 2D or 3D) and models were employed. The segmentation outcomes were quantitatively assessed using the Agatston score to enhance clarity and reduce subjective bias:

$$\text{Agatston Score} = \sum_{\text{slices}} (\text{Area} \times \text{Density Factor}) \tag{5}$$

This integrated approach uses advanced image processing techniques and quantitative metrics to improve plaque segmentation accuracy in medical imaging while addressing technical challenges and improving the reliability of the results.

4 Experimental Design

4.1 Dataset

The dataset originates from the orCaScore challenge [26]. This study used a comprehensive dataset comprising cardiac CT-based calcium scoring images, collected from various computed tomography (CT) scanners of different manufacturers, located in four separate hospitals. Each patient's scan included two types of images: non-contrast and contrast-enhanced CTA. For the purpose of the following work only non-contrast images were used. The training set included 32 patients, while the testing set comprised 40 individuals, ensuring a robust evaluation framework. The benchmark for non-contrast CT scans was established through a comprehensive analysis of all provided Cardiac CT (CSCT) scans. This analysis was carried out by two expert observers: a radiologist with 12 years of experience and more than 5000 CSCT scans evaluated (TL) and a research physician with five years of experience and more than 1,000 CSCT scans evaluated [26]. Following vendor guidelines, image pixels with values greater than 130 Hounsfield units were highlighted. Calcifications were manually classified according to their location within the left anterior descending (LAD), left circumflex (LCx), or right coronary artery (RCA). In this challenge, calcifications in the left main coronary artery were classified as LAD-related. The images annotated with reference labels had calcification labels of 1 (LAD), 2 (LCX), or 3 (RCA). Any lesions where the expert opinions differed were subsequently reviewed and resolved by consensus, establishing the reference standard.

4.2 Preprocessing

For two-dimensional images, resizing was performed to achieve consistent dimensions, and alignment to the RAS coordinate system standardized their orientation. Intensity values were scaled between –200 and 1300 HU. This approach allowed peripheral areas, such as the myocardium, to serve as additional reference points, using the presence of water as a feature of diminished influence rather than completely eliminating it. Finally, label binarization enabled for precise plaque detection in the coronary arteries. This process was specifically designed to highlight the presence of plaques, focusing on their detection rather than their exact location within the arterial segments. The preprocessing of three-dimensional data focused on maintaining consistent spatial resolution and aligning with the RAS coordinate system. The other steps were the same as for the 2 D input data type.

4.3 Data Augmentation

In 3D segmentation, volumetric data were sampled in cubes of shape (96, 96, 96) with a ratio of 4:1 for positive to negative samples, focusing on pathological regions while maintaining contextual understanding. Intensity shifts and contrast adjustments improved model robustness to minor variations. For 2D segmentation, careful sampling addressed the challenge of displaying unbalanced plaques in slices within volumes. The plaques, being small structures, occupy only a few slices, and random sampling from volumes with a minimum of 50 slices would result in an imbalance, predominantly selecting non-pathological samples and underrepresenting plaques. To address this, raw labels were prepared exclusively for plaques, excluding other structures such as the left and right atrium of the heart. This ensured that the sampled slices contained plaque-containing regions. The slices were resized uniformly without altering the spacing, optimizing the efficiency of model training.

5 Training and Evaluation

Training. The primary objective of this study was to evaluate the performance of various neural network architectures for calcium scoring in cardiac CT images. We tested several models including RaUNet2D, RaUNet3D, UNet2D, UNet3D, and UNETR2D. Each model was trained with a consistent set of hyperparameters to ensure comparability.

RaUNet2D and RaUNet3D [18]: These models use an Attention Residual U-Net architecture for 2D and 3D data respectively. The network comprises channels (16, 32, 64, 128, 256) and strides (2, 2, 2, 2). Batch normalization [9] and dropout layers [20] are used to prevent overfitting. The models utilize the Dice Cross-Entropy Loss (DiceCELoss) [5] function and Dice Metric [22] for evaluation. The optimizer employed is AdamW with a learning rate of 1×10^{-4}.

UNet2D and UNet3D [19]: These models are based on the standard UNet architecture adapted for 2D and 3D data. They feature similar channel and stride configurations as RaUNet2D and RaUNet3D, with 2 residual units and batch normalization.

UNETR2D [8]: This model is based on the UNETR architecture for 2D data, with feature size 8, hidden size 384, MLP dimension 1536, and 6 heads.

All models were trained using mixed-precision training [17] (precision: 16mixed) on a GPU. The training lasted for a maximum of 1600 epochs with early stopping based on validation loss. The batch size was set to 6 for 2D models and 1 for 3D models due to memory constraints. AdamW [15] Optimizer with a learning rate of 1×10^{-4} was used for all models.

The selected architectures were chosen to investigate the trade-off between computational efficiency and architectural complexity in the context of cardiac CT calcium scoring. UNet2D and UNet3D serve as reliable baselines, known for their effectiveness in medical imaging, while RaUNet2D and RaUNet3D leverage attention mechanisms and residual connections for enhanced precision and gradient flow. UNETR2D introduces transformer-based global context learning, enabling the study to assess the impact of cutting-edge methodologies. By including both 2 D and 3 D variants, the study explores the trade-offs between computational efficiency and the differences in accuracy of predictions based on volumetric context.

Evaluation. The main criterion for assessing model performance and the influence of various input data types was the use of the Agatston score. By comparing the predicted Agatston scores with the actual scores, we could assess not only the accuracy of the predictions, but also their practical relevance in a clinical context. This ensured that the model output was accurate and meaningful for medical diagnosis and treatment planning. Another important factor in the evaluation process was the percentage of plaque coverage detected by the model compared to the actual data. This metric provided insight into the models ability to accurately identify and outline plaques within the imaging data. A high percentage of coverage indicated that the model effectively captured the extent of plaque presence, which is crucial for reliable diagnosis and subsequent medical interventions. The final evaluation was performed per patient, allowing for a detailed evaluation of the model performance between individual patients. This approach was essential to understand the variability in model performance due to patient-specific factors and different imaging conditions. By analyzing the results at the patient level, we could identify any inconsistencies or biases in the model predictions, ensuring a robust evaluation process. In addition, the impact of different image sources on model performance was thoroughly analyzed. This involved evaluating how variations in image acquisition conditions, such as different imaging devices, protocols, and settings, affected model accuracy. Understanding this impact was vital to ensure the generalizability of the model in various clinical settings and imaging conditions. In general, these comprehensive validation criteria ensured that the model was rigorously evaluated not only for its predictive accuracy but also for its practical utility in real-world medical applications.

6 Results

6.1 Quantitative Comparison

This section compares the 2D and 3D methods. Table 1 presents the Dice scores across different channels, illustrating their overall performance, especially in segmenting complex structures such as plaques. Using 2D input within RaUNet highlights its effectiveness in plaque segmentation, showcasing its methodological strength and practical importance in medical imaging applications. For a detailed comparative analysis, Figs. 1 and 2 present comprehensive Agatston results obtained from the validation set.

Table 1. Segmentation results obtained by 2D and 3D architectures. The approaches were evaluated according to Dice score and absolute errors (mean ± st. Dev.) in measured Agatston score and plaque volume.

Input Dim. Model	Dice (↑)	Agatston (↓)	Plaque (↓)	
2D	UNet	0.51	166.98 ± 248.76	158.65 ± 201.59
	UNeTR	0.41	229.38 ± 319.95	200.78 ± 280.89
	RAUNet	0.56	21.04 ± 52.95	22.74 ± 48.88
3D	UNet	0.84	45.49 ± 60.46	40.69 ± 59.69
	RAUNet	0.87	52.14 ± 120.36	54.44 ± 127.02

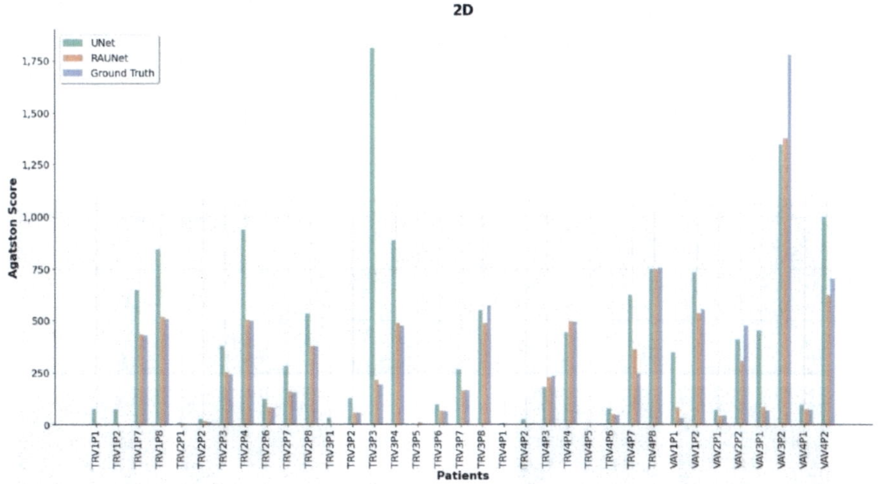

Fig. 1. Comparison of Agatston results of different methods for 2D model variants. Labels refer to samples from the OrCaScore dataset.

6.2 Qualitative Comparison

The upcoming subsection offers a qualitative analysis of the activation maps shown in Fig. 4 and the model predictions displayed in Fig. 3. These maps emphasize the areas where the model concentrates during prediction. Each illustration arranges the activation maps in a left-to-right order, corresponding to the channels that depict plaques, aortas, and other anatomical structures. Rather than solely extracting the highest value from the softmax outputs, the activation maps were generated by inspecting the results of individual channels. This method provides a more detailed comprehension of the model's attention across various anatomical features. Through the inspection of activation maps for each channel, we obtained a deeper understanding of the models' decision-making mechanisms, thereby enhancing explainability. Consequently, we observed that enriching contextual information by integrating additional anatomical structures boosts model performance, albeit at the cost of increased complexity with 3D data.

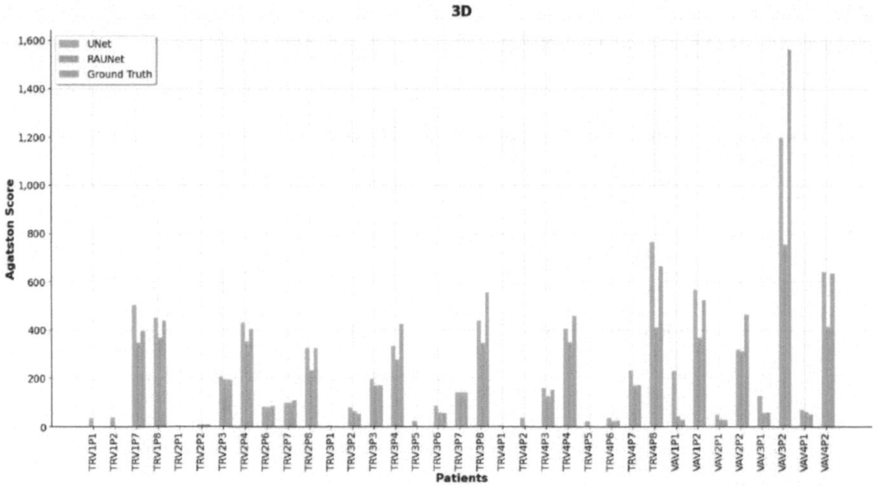

Fig. 2. Comparison of Agatston results of different methods for 3D model variants. Labels refer to samples from the OrCaScore dataset.

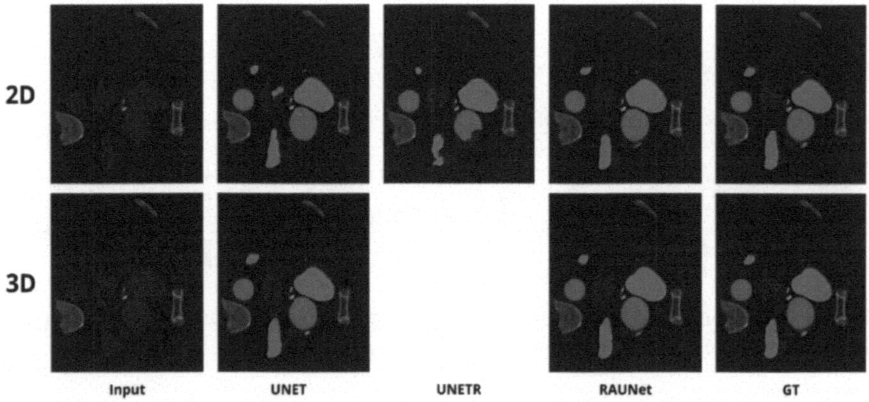

Fig. 3. Segmentation results comparision for different models. Orange represents plaques, purple represents anatomical structures, such as the right atrium of the heart, the left atrium, the myocardium, the left ventricle, and the right ventricle and color green represents aorta.

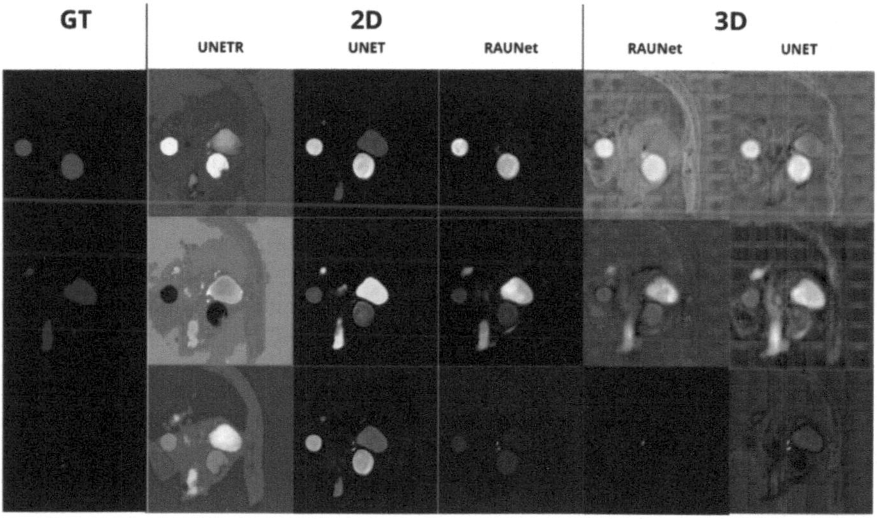

Fig. 4. Segmentation results comparision for different models. Orange labels depict plaques, purple delineates key anatomical structures including the right atrium, left atrium, myocardium, left ventricle, and right ventricle, while green highlights the aorta. Presented are varied activation maps corresponding to distinct channels, showcasing the aorta, anatomical structures, and plaques from top to bottom. From up to down presented are aorta, anatomical structures mentioned earlier and plaques.

7 Conclusions

Developing robust machine learning models for plaque detection and calcium scoring in CT cardiac imaging is challenging, particularly because of the substantial inter-slice distances. This can cause the model to learn redundant features, and as a result, there is a risk that it might ignore critical areas, especially when segmenting small objects appearing in only a few slices. Employing large volumes along the z-axis can lead to imprecise feature representation.

Simple 2D models lose crucial spatial structure, mistaking noise for regions of interest. In 3D data, high activation values of large structures such as the myocardium overshadow the gradients of plaques, causing discontinuities and inaccurate representations. This is worsened by low spatial resolution and limited training images. Background noise also significantly impacts model performance, with 2D models generally minimizing its influence better than 3D models due to less sparse matrix generation. However, for segmenting smaller objects, a hybrid approach can achieve better results. Quantitative and qualitative analyzes indicate that 3D models are not ideal for low-spatial resolution segmentation tasks.

References

1. Abdelrahman, K., et al.: Artificial intelligence in coronary artery calcium scoring detection and quantification. Diagnostics **14**(2), 125, January 2024. https://doi.org/10.3390/diagnostics14020125

2. Agatston, A.S., Janowitz, W.R., Hildner, F.J., Zusmer, N.R., Viamonte, M., Detrano, R.: Quantification of coronary artery calcium using ultrafast computed tomography. J. Am. Coll. Cardiol. **15**(4), 827832 (1990). https://doi.org/10.1016/0735-1097(90)90282-T. https://www.sciencedirect.com/science/article/pii/073510979090282T
3. Auddy, A., Zou, H., Rad, K.R., Maleki, A.: Approximate leave-one-out cross validation for regression with ℓ_1 regularizers (extended version) (2023). https://doi.org/10.48550/ARXIV.2310.17629
4. van der Bijl, N., et al.: Assessment of agatston coronary artery calcium score using contrast-enhanced CT coronary angiography. Am. J. Roentgenol. **195**(6), 1299–1305, December 2010. https://doi.org/10.2214/ajr.09.3734
5. Galdran, A., Carneiro, G., Ballester, M.A.G.: On the optimal combination of crossentropy and soft dice losses for lesion segmentation with out-of-distribution robustness (2022). https://doi.org/10.48550/ARXIV.2209.06078
6. Greenland, P., et al.: ACCF/AHA 2007 clinical expert consensus document on coronary artery calcium scoring by computed tomography in global cardiovascular risk assessment and in evaluation of patients with chest pain: a report of the American college of cardiology foundation clinical expert consensus task force. J. Am. Coll. Cardiol. **49**(3), 378–402 (2007). https://doi.org/10.1016/j.jacc.2006.10.001. https://www.sciencedirect.com/science/article/pii/S0735109706024636
7. Gupta, A., et al.: Coronary artery calcium scoring: current status and future directions. RadioGraphics **42**(4), 947–967, July 2022. https://doi.org/10.1148/rg.210122
8. Hatamizadeh, A., et al.: UNETR: transformers for 3d medical image segmentation (2021). https://doi.org/10.48550/ARXIV.2103.10504. https://arxiv.org/abs/2103.10504
9. Ioffe, S., Szegedy, C.: Batch normalization: accelerating deep network training by reducing internal covariate shift (2015). https://doi.org/10.48550/ARXIV.1502.03167. https://arxiv.org/abs/1502.03167
10. Kang, H.W., et al.: Evaluation of fully automated commercial software for Agatston calcium scoring on non-ECG-gated low-dose chest CT with different slice thickness. Euro. Radiol. **33**(3)
11. Kim, K.P., Einstein, A.J., De Gonzalez, A.B.: Coronary artery calcification screening: estimated radiation dose and cancer risk. Arch. Int. Med. **169**(13), 1188–1194 (2009)
12. Lee, H., et al.: Machine learning and coronary artery calcium scoring. Curr. Cardiol. Reports **22**(9), July 2020. https://doi.org/10.1007/s11886-020-01337-7
13. Lee, J., et al.: Prediction of coronary artery calcium score using machine learning in a healthy population. J. Personal. Med. **10**(3), 96, August 2020. https://doi.org/10.3390/jpm10030096
14. Lin, A., et al.: Deep learning-enabled coronary CT angiography for plaque and stenosis quantification and cardiac risk prediction: an international multicentre study. Lancet Dig. Health **4**(4), e256–e265 (2022)
15. Loshchilov, I., Hutter, F.: Decoupled weight decay regularization (2019). https://arxiv.org/abs/1711.05101
16. Martin, S.S., et al.: Evaluation of a deep learning–based automated CT coronary artery calcium scoring algorithm. JACC Cardiovas. Imaging **13**(2), 524–526, February 2020. https://doi.org/10.1016/j.jcmg.2019.09.015
17. Micikevicius, P., et al.: Mixed precision training (2017). https://doi.org/10.48550/ARXIV.1710.03740. https://arxiv.org/abs/1710.03740
18. Ni, Z.L., et al.: RAUNet: residual attention u-net for semantic segmentation of cataract surgical instruments (2019). https://doi.org/10.48550/ARXIV.1909.10360. https://arxiv.org/abs/1909.10360
19. Ronneberger, O., Fischer, P., Brox, T.: U-net: Convolutional networks for biomedical image segmentation (2015). https://doi.org/10.48550/ARXIV.1505.04597, https://arxiv.org/abs/1505.04597

20. Srivastava, N., Hinton, G., Krizhevsky, A., Sutskever, I., Salakhutdinov, R.: Dropout: a simple way to prevent neural networks from overfitting. J. Mach. Learn. Res. **15**(56), 1929–1958 (2014). http://jmlr.org/papers/v15/srivastava14a.html
21. Suh, Y.J., et al.: Fully automatic coronary calcium scoring in non-ECG-gated low-dose chest CT: comparison with ECG-gated cardiac CT. Euro. Radiol. **33**(2), 1254–1265, September 2022. https://doi.org/10.1007/s00330-022-09117-3
22. Taha, A.A., Hanbury, A.: Metrics for evaluating 3d medical image segmentation: analysis, selection, and tool. BMC Med. Imaging **15**(1), August 2015. https://doi.org/10.1186/s12880-015-0068-x
23. Wang, L., et al.: PlaqueNet: deep learning enabled coronary artery plaque segmentation from coronary computed tomography angiography. Visual Comput. Indust. Biomed. Art **7**(1), March 2024. https://doi.org/10.1186/s42492-024-00157-8
24. Wang, W., et al.: Coronary artery calcium score quantification using a deep-learning algorithm. Clin. Radiol. **75**(3), 237.e11–237.e16 (2020). https://doi.org/10.1016/j.crad.2019.10.012. https://www.sciencedirect.com/science/article/pii/S0009926019306099
25. Wasserthal, J., et al.: Totalseg - mentator: robust segmentation of 104 anatomic structures in CT images. Radiol. Art. Intell. **5**(5), September 2023. https://doi.org/10.1148/ryai.230024
26. Wolterink, J.M., et al.: An evaluation of automatic coronary artery calcium scoring methods with cardiac CT using the orcascore framework. Med. Phys. **43**(5), 2361–2373 (2016)
27. Yu, J., et al.: Automated total and vessel-specific coronary artery calcium (CAC) quantification on chest CT: direct comparison with CAC scoring on non-contrast cardiac CT. BMC Med. Imaging **22**(1), October 2022. https://doi.org/10.1186/s12880-022-00907-1

Towards Objective Assessment of the Accuracy of Coronary Artery Segmentation

Jakub Pałachniak[1,2](✉), Paweł Luniak[1], Maciej Zamorski[1], Mateusz Kierepka[1], and Karol Miller[3]

[1] Hemolens Diagnostics Sp. z o.o., Legnicka 48G, 54-202 Wroclaw, Poland
{jakub.palachniak,pawel.luniak}@hemolens.com
[2] Silesian University of Technology, Akademicka 2A, 44-100 Gliwice, Poland
[3] Intelligent Systems for Medicine Laboratory, The University of Western Australia, Perth WA 6009, Australia

Abstract. In recent years the interest in computational fluid dynamics (CFD)-based analysis of blood flow in coronary arteries has intensified, largely motivated by the possibility of non-invasive estimation of fractional flow reserve (FFR). FFR is important in the management of heart disease as it indicates whether there is insufficient blood flow entering the myocardium that would indicate a surgical intervention to restore sufficient flow. There have been various CFD-based modelling methodologies proposed in the literature with all of them crucially depending on the fidelity of coronary vasculature reconstruction from medical images - a process known as image segmentation. In this contribution, we describe an evaluation methodology of coronary artery segmentations developed by Hemolens Diagnostics and the objective assessment of its accuracy based on B-Spline transform. The results indicate that this methodology is an effective way to evaluate the model's robustness to distortions.

Keywords: Coronary artery segmentation · Machine Learning · FFR

1 Introduction

Coronary artery disease (CAD) is one of the leading causes of death in the European Union and the USA [2,14]. Invasive coronary angiography is considered the gold standard for detecting and assessing the severity of coronary artery disease. However, as an invasive procedure, it poses risks of adverse events related to the procedure itself. When non-invasive stress imaging is not feasible, non-diagnostic, or unavailable, fractional flow reserve (FFR) or coronary flow reserve measurements become crucial during diagnostic coronary angiography. Research indicates that it is safe to defer percutaneous coronary intervention (PCI) or coronary artery bypass grafting (CABG) in patients with an FFR above 0.80 [4].

J. Pałachniak and P. Luniak—Contributed equally to this work.

M. Kobielarz et al. (Eds.): MICCAI 2024, LNBE, pp. 50–61, 2025.
https://doi.org/10.1007/978-3-031-94128-3_6

FFR assessment is particularly useful for understanding the functional consequences of moderate coronary stenoses. Additionally, FFR-guided PCI combined with medical therapy has been shown to reduce the necessity for urgent revascularization compared to the best available medical therapy alone [2,17]. The significance of FFR lies in its ability to provide a precise physiological assessment of the severity of coronary artery stenoses. Unlike traditional angiography, which primarily offers anatomical information, FFR evaluates the functional impact of the coronary lesion on blood flow. This functional assessment is vital because it directly correlates with the likelihood of ischemia [11]. While FFR is a valuable tool, its invasive nature presents certain risks and limitations, prompting the exploration of non-invasive alternatives. Computational Fluid Dynamics (CFD) has emerged as a promising technology in this regard [15]. CFD utilizes mathematical models and algorithms to simulate blood flow and pressure within the coronary arteries, leveraging imaging data such as computed tomography (CT) scans.

The application of CFD in cardiovascular diagnostics allows for the calculation of a non-invasive FFR (FFR CT), which can provide similar functional information about coronary artery lesions without the need for catheterization. This approach involves the reconstruction of coronary anatomy from CT images, followed by the simulation of blood flow dynamics to assess the impact of arterial narrowing on blood flow and pressure [2]. The advancement of CFD technology represents a significant step forward in the non-invasive diagnosis and management of heart diseases. By integrating computational models with medical imaging, CFD enables a detailed and patient-specific analysis of coronary physiology, potentially transforming the landscape of cardiovascular diagnostics [3].

For widespread deployment of CFD-based technology for diagnostics and management of coronary artery disease it is necessary to develop robust and reliable methods of coronary vessel segmentation and be able to objectively assess segmentation accuracy.

Recent years have seen a surge in the application of deep learning techniques for coronary vessel segmentation, driven by the success of convolutional neural networks (CNNs) [5,7]. The U-Net architecture, introduced by Ronneberger in 2015 [13], has become a benchmark for medical image segmentation due to its ability to capture both local and global context through its encoder-decoder structure. U-Net and its variants, such as U-Net++ [18] and Attention U-Net [8], have demonstrated impressive performance in segmenting coronary vessels from computed tomography angiography (CTA) images. T-Net incorporates multiple small encoder-decoders within each block to capture feature maps of various sizes, enhancing segmentation accuracy. This approach has shown improvement over traditional U-Net models, achieving higher Dice coefficient in coronary segmentation [10].

However, the medical imaging is often distorted, e.g. due to the movement of the patient, reconstruction error, foreign object or phenotype of a disease like calcium blooming [1,7]. Therefore it is crucial to assess the quality of automatic segmentation methods not only on clean, sharp images but also quantify the performance degradation coming from distorsion in the image. In this work, we outline an evaluation methodology for coronary artery segmentation using B-Spline transform and assess its accuracy. To the best of our knowledge, this is the first instance where this methodological outline has been both developed and described in the literature.

2 Methods

2.1 Patient Data

For the experiments we used a subset containing 209 training and 23 test anonymised contrast-enhanced computed tomography angiography (CCTA) images from a dataset of patients who gave their informed consent before their participation in the study, recruited as a part of a bioethics committee approved clinical trial (ClinicalTrials.gov ID NCT04777513). Images were obtained with a standard CCTA field of view, ensuring comprehensive coverage of the coronary arteries, including the left main, left anterior descending, left circumflex, and right coronary arteries, as well as adjacent cardiac structures. The key characteristics of test subset are summarized in Table 1.

Table 1. Participant-level characteristics of the study cohort. Values provided as mean ± std. dev. or as a percentage.

Variable	Abbr.	Units	Value
Age	–	year	67.2 ±9.4
Height	–	cm	172.4 ± 8.5
Weight	–	kg	82.3 ± 13.0
Heart Rate	HR	/min	65.6 ± 9.3
Diastolic BP	DIA	mmHg	81.6 ± 11.1
Systolic BP	SYS	mmHg	143.6 ± 17.9
Hematocrit	HCT	%	40.4 ± 2.6
Body Mass Index	BMI	kg/m^2	27.7 ± 4.0
Male participants	–	%	76.0
Hypertension	–	%	56.0
Current smoker	–	%	12.0
AARR drugs	–	%	20.0
BBLOCK drugs	–	%	84.0
ACE drugs	–	%	44.0

2.2 Segmentation Method

For the purposes of this work, we use a custom UNet-based architecture trained on a subset of private data consisting of 209 patients. The data was split into training and validation sets with an 80/20 ratio. The model was trained until convergence, using the Adam optimizer, ensuring that the model achieved optimal performance on the validation set. Both segmentations and B-Spline transform evaluation were done with Python 3.9.17 and PyTorch 2.0.1.

2.3 Method for Objective Evaluation of Segmentation Accuracy

In most published studies evaluating segmentation accuracy, manual segmentation performed by an expert is considered the so-called *ground truth*. The differences between segmentations produced by a proposed automatic method and this manually prepared ground truth are considered segmentation errors. However, significant evidence shows that expert's manual segmentation cannot be fully trusted due to inter- and intra-observer variability [9,12]. Inter-observer variability refers to the differences in segmentation results when different experts segment the same images. Each expert may have a unique interpretation of the image features, leading to variations in the boundaries and regions identified. This variability can be influenced by several factors, including the expert's experience, understanding of the anatomy, and the specific segmentation protocols they follow. Intra-observer variability occurs when the same expert produces different segmentation results when performing the task multiple times on the same set of images. This inconsistency can arise due to factors such as fatigue, changes in focus, or subjective interpretation of ambiguous regions within the image. Even slight variations in the manual tracing process can lead to significant differences in the final segmented regions. In the absence of the ground truth, objective evaluation of segmentation accuracy presents itself as a significant challenge.

To address this problem, we propose the following 5 steps protocol:

1: Generate B-Spline Transformation. This step involves creating a B-Spline transformation [16], denoted as T. B-Splines are piecewise polynomial functions that can represent complex, smooth shapes. In image processing, B-Spline transformations are used for non-rigid (deformable) transformations that smoothly warp an image. This is accomplished by defining a grid of control points over the image and applying polynomial interpolation between these points to achieve the transformation. The control points can be manipulated to create the desired deformation of the image.

B-Spline transformations are generated using the BSplineTransformInitializer function from the SimpleITK library. This function initializes the B-Spline transform by setting up a grid of control points over the given image. In our setup the grid consists of $10 \times 10 \times 10$ control points evenly distributed across the three dimensions of the image. After the grid is established, we introduce random displacements to these control points. The displacement for each control point is taken from a normal distribution, with the standard deviation (st. dev.) of the distribution controlled by the parameter σ (sigma). A larger σ results in more significant displacements, leading to greater deformation of the image. The applied displacements are constrained to prevent control points from overlapping or causing unrealistic folding of the image, thereby maintaining the anatomical coherence of the structures.

To ensure that the transformation is smooth and realistic, we set the displacements of control points located on the boundaries of the grid to zeros. This prevents unrealistic warping at the edges of the image. The combination of controlled random displacements and boundary conditions allows us to generate a variety of realistic B-Spline transformations for use in the experiments.

2: Make predictions on original CCTA image $\boldsymbol{x}$ Using our segmentation model, predictions are made on the original CCTA image x. The output, $\hat{y}$, represents the segmentation of the coronary arteries within the image.

3: Transform Original CCTA Image. The original CCTA image x is transformed with the B-Spline transformation T, resulting in a new, deformed image x_T. The same segmentation model is then applied to this transformed image to obtain predictions $\hat{y}_{x_T}$.

4: Transform Segmentations. The segmentation $\hat{y}$ obtained from the original CCTA image is transformed using the same B-Spline transformation T to produce $\hat{y}_T$.

5: Compare $\hat{\boldsymbol{y}}_T$ and $\hat{\boldsymbol{y}}_{x_T}$. The final step is to compare the transformed segmentation $\hat{y}_T$ with the segmentation obtained directly from the transformed image $\hat{y}_{x_T}$.

This protocol is repeated for each examined volume with increasing σ values, which define the standard deviation of the random displacement of control points in the grid during the creation of the B-Spline. This allows us to control the level of image distortion. The results of this protocol are pairs of meshes representing segmentations for a given volume and a given σ value - the original mesh distorted by the transformation T and the mesh obtained by segmenting the image distorted by the same transformation T. Comparing these two meshes allows us to assess the model's robustness to image artifacts, lower quality scans, and anatomical differences among patients. To compare the meshes, we use two approaches: 3D - comparing the entire meshes in 3D space, and 2D - comparing specific cross-sections of individual vessels, described in Sect. 2.4. While B-Spline transforms may potentially create unrealistic deformations, especially in soft tissue, our use of a relatively small number of control points ($10\times10\times10$) keeps the transformations globally smooth. This setup primarily adjusts the position of small structures like blood vessels without significantly altering their shape. The applied deformations may mimic clinical variations in vessel shape and wall thickness due to patient-specific differences [6].

2.4 Metrics

3D Evaluation. For 3D evaluation Dice coefficient is computed to indicate the overall 3D overlap of obtained meshes. The Dice coefficient, a measure of similarity, is widely used in evaluating segmentation quality in machine learning and computer vision. Its value ranges from 0.0 to 1.0, where 1.0 indicates perfect alignment between compared regions and 0 indicates no overlap at all. The Dice coefficient between samples A and B is defined as:

$$\text{Dice}(A, B) = \frac{2\, A \cap B|}{|A| + |B|} \tag{1}$$

where:

- $|A \cap B|$ is the volume of the intersection of the two meshes A and B,
- $|A|$ is the volume of mesh A,
- $|B|$ is the volume of mesh B.

The second metric we use to evaluate meshes is Hausdorff Distance. The Hausdorff Distance is a metric used to evaluate the similarity between two 3D segmentations by determining the maximum distance between the surfaces of the segmented objects. It involves calculating the greatest distance from a point on the surface of one segmentation to the nearest point on the surface of the other segmentation. This metric is very sensitive to differences and provides a strict measure of how well two segmentations match.

The Hausdorff Distance $H(A, B)$ between two sets of points A and B can be defined as:

$$H(A, B) = \max(h(A, B), h(B, A)), \tag{2}$$

where $h(A, B)$ is the directed Hausdorff distance from set A to set B, given by:

$$h(A, B) = \max_{a \in A} \min_{b \in B} \|a - b\|. \tag{3}$$

Here, $\|a-b\|$ denotes the Euclidean distance between points a and b. This formula essentially captures the maximum of the minimum distances from each point in one set to the nearest point in the other set, thereby providing a measure of the greatest mismatch between the two segmentations.

2D Evaluation. In order to evaluate meshes in two dimensions, we calculate the Hausdorff Distance, the Dice coefficient, and the cross-sectional area defined by a plane perpendicular to the direction of blood flow and intersecting a point located on the centerline of the vessel. These metrics are determined for cross-sections along the entire length of the vessel, covering each of its branches for both segmentations.

3 Results

From the cohort of 232 patients we set aside 23 images not seen during the training of the segmentation model that were used for evaluation purposes. For each volume, we generated 20 B-Spline transformations with standard deviations of the control points displacement ranging from 1 mm to 20 mm in steps of 1 mm. Dice coefficient and Hausdorff distance were calculated between the resulting pairs of meshes. Sample slices and corresponding meshes are shown in Fig. 1. In Fig. 2 we present a comparison between two produced meshes generated during steps 2 and 4 of the process presented in Sect. 2.3.

As shown in Fig. 3, there is a slight but steady decrease in the Dice coefficient as distortion increases. Up to a distortion level of 12 mm, the Dice coefficient remains above 0.9. Further increasing the distortion beyond 12 mm results in a faster decline in the Dice coefficient, but even at the maximum σ of 20 mm, the coefficient does not drop below 0.8 for any of the meshes.

The Hausdorff distance metric shown in Fig. 3 exhibits a similar trend to Dice coefficient. Within the distortion range of 1 to 12 mm, there is an approximately linear, slight decrease in segmentation quality, resulting in an increase in the metric value from below 0.15 mm to above 0.20 mm. Beyond a σ of 12 mm, there

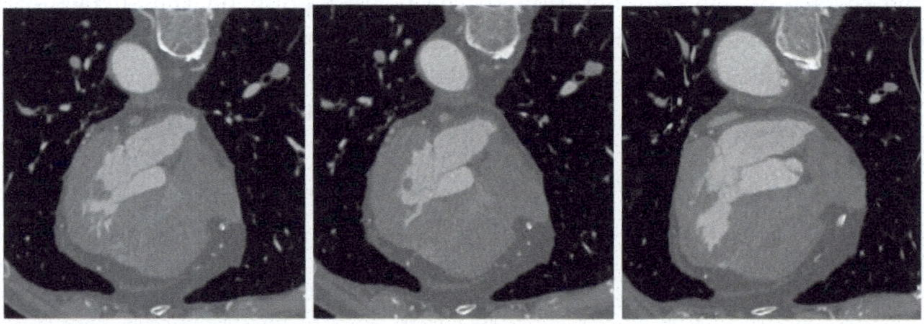

(a) $\sigma = 1$ mm (b) $\sigma = 5$ mm (c) $\sigma = 20$ mm

(d) $\sigma = 1$ mm (e) $\sigma = 5$ mm (f) $\sigma = 20$ mm

Fig. 1. Sample slices from cardiac CCTA image distorted with B-Splines generated with different σ values (top row) with corresponding coronary arteries segmentation (bottom row).

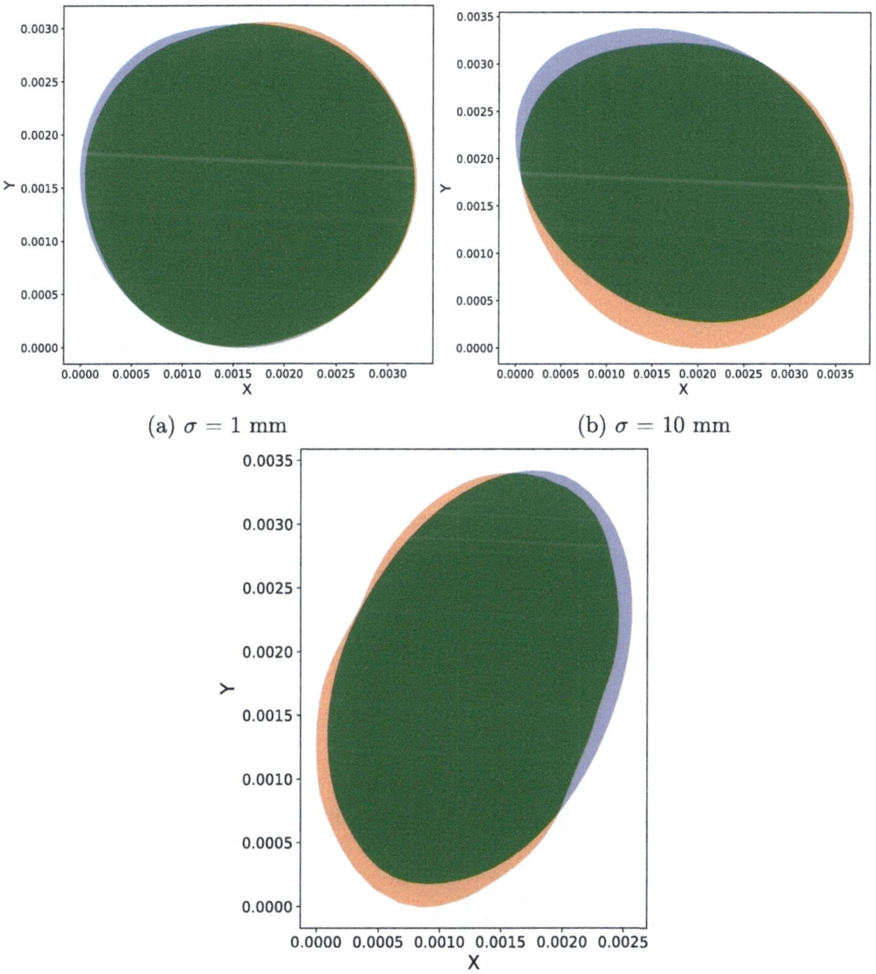

(a) $\sigma = 1$ mm (b) $\sigma = 10$ mm

(c) $\sigma = 20$ mm

Fig. 2. Sample cross-sections of meshes distorted with B-Splines generated with different σ values. The green area represents the common part of both segmentations, the blue area represents the segmentation transformed with B-Spline, and the red area represents the segmentation made on the transformed volume. It is noticeable that the more distorted the input, the more deformed the segmented vessel lumen is-becoming increasingly oval.

is an accelerated decline in the metric. The Fig. 4 shows that the 95th percentile of the Hausdorff distance is below 0.4 mm.

For each mesh transformed using B-Spline, between 250 and 400 cross-sections were obtained, depending on the length of the vessels and the number of branches. For each cross-section, the corresponding cross-section was

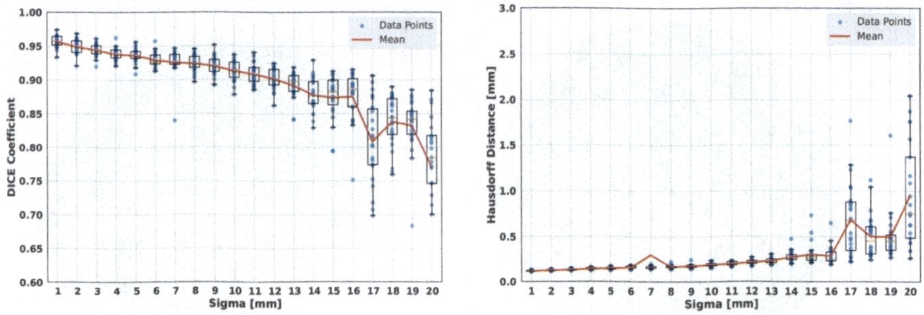

Fig. 3. Increasing discrepancies between the transformed mesh and the segmentation shown as Dice score lowering (left) and Hausdorff distance rising (right) with higher st. dev. σ of the displacement of grid nodes used for deforming the image and mesh.

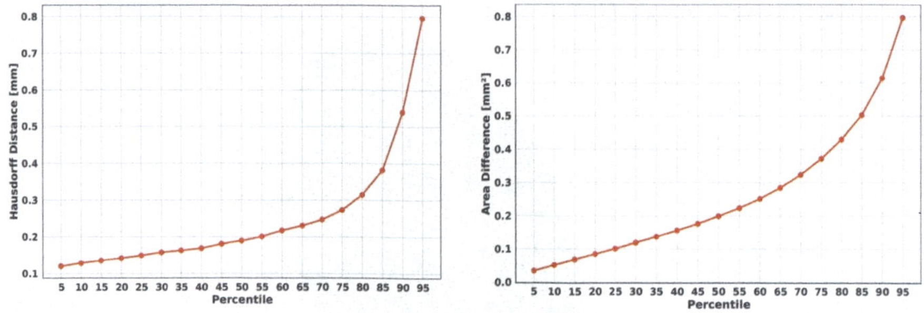

Fig. 4. Comparison of Hausdorff Distance percentiles for 3D segmentations (left) and area difference percentiles for cross-sections (right) as a function of σ.

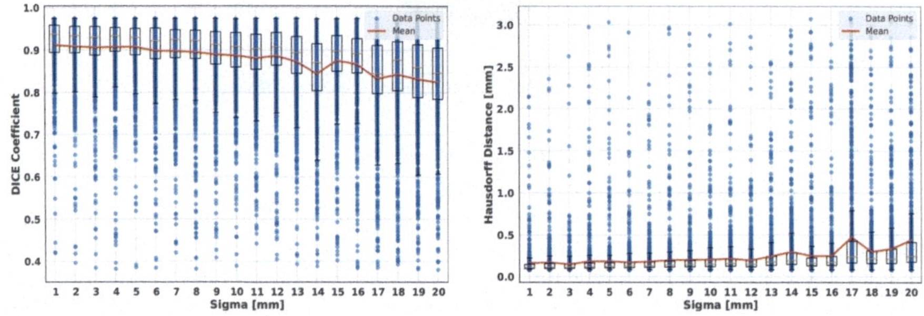

Fig. 5. Comparison of Dice coefficient (left) and Hausdorff Distance (right) for cross-sections as a function of σ. The overlap of cross-sections determined using the Dice coefficient decreases as σ increases, indicating lower segmentation quality for stronger distortions. Similarly, the Hausdorff Distance increases with higher σ values, indicating greater discrepancies between the transformed mesh and the segmented cross-sections.

identified in the mesh resulting from the segmentation of the transformed image. The comparison of the Dice coefficient for the cross-sections is shown in Fig. 5.

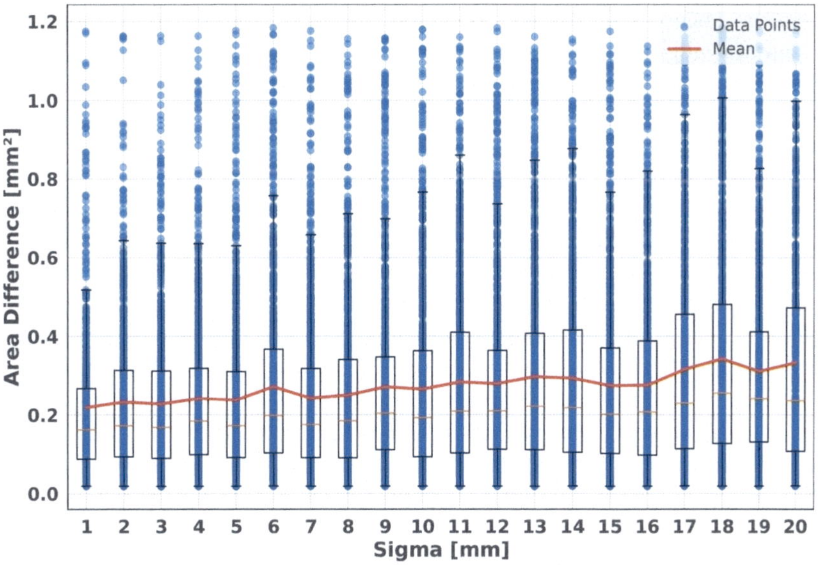

Fig. 6. The area differences of cross-sections increases by 0.15 mm^2 with increasing σ, indicating lower segmentation quality for stronger distortions.

The Dice coefficient for cross-sections exhibits a similar trend to the Dice coefficient for meshes. Up to a σ value of 13 mm, the coefficient's mean decreases linearly without falling below 0.85. Beyond a σ value of 12 mm, a slight decline in the metric is observed. The Hausdorff distance calculated for pairs of cross-sections is shown in Fig. 5. This metric slightly increases with the increase in σ values, but similarly to the meshes' metrics, the average value do not exceeds 0.5 mm until σ=16 mm.

For cross-sections, the differences in the cross-sectional area were also calculated - this is the most important parameter from the perspective of blood flow analysis. These differences are shown in Fig. 6. The average difference in cross-sectional area, regardless of the σ value, ranges between 0.2 mm^2 and 0.4 mm^2. With the increase in distortion, a steady but slight increase in the metric is observed. The Fig. 4 shows that the 95th percentile of the area difference is below 0.8 mm^2 with 90th percentile being under 0.6 mm^2.

4 Discussion and Conclusion

Our method for evaluating the accuracy of coronary artery segmentation using B-Spline transformations has proven to be effective in assessing the robustness of segmentation models. By systematically introducing controlled distortions to the original CCTA images, we were able to test the segmentation model's performance under various degrees of image deformation. The use of metrics such as the Dice coefficient and Hausdorff distance allowed us to quantify the precision and alignment of the segmentations.

The results indicate that the segmentation model maintains high accuracy up to high levels of distortion ($\sigma = 12\,\text{mm}$), as shown by Dice coefficients above 0.9 and 0.85 for 3D mesh similarity and cross-sections areas respectively as well as Hausdorff distances below 0.25 mm for both those metrics.

Overall, we show that our evaluation method using B-Spline transformations is effective in testing the robustness and accuracy of coronary artery segmentation models and in the absence of distorted ground truth may indicate that the model performs well under most of the typical clinical conditions. Future work will focus on evaluating the influence of segmentation differences on pressure drop calculations in CFD-based analyses such as non-invasive FFR predictions and exploring other types of image distortions to further enhance the model's robustness and clinical applicability. We acknowledge that future refinements could include constraints to prevent excessive changes in vessel morphology, enhancing the accuracy and realism of the ground truth data while better simulating patient variability.

References

1. Barrett, J.F., Keat, N.: Artifacts in ct: recognition and avoidance. Radiographics **24**(6), 1679–1691 (2004)
2. Boot, I.W., Planken, R.N., den Hartog, A.W., Vrijhoef, H.J.: Early-stage health technology assessment of fractional flow reserve coronary computed tomography versus standard diagnostics in patients with stable chest pain in the netherlands. PLoS ONE **19**(6), e0305189 (2024)
3. Coenen, A., et al.: Diagnostic accuracy of a machine-learning approach to coronary computed tomographic angiography–based fractional flow reserve: result from the machine consortium. Circulation: Cardiovascular Imaging **11**(6), e007217 (2018)
4. Collet, C., Onuma, Y., Sonck, J., Asano, T., Vandeloo, B., Kornowski, R., Tu, S., Westra, J., Holm, N.R., Xu, B., et al.: Diagnostic performance of angiography-derived fractional flow reserve: a systematic review and bayesian meta-analysis. Eur. Heart J. **39**(35), 3314–3321 (2018)
5. Dong, C., Xu, S., Dai, D., Zhang, Y., Zhang, C., Li, Z.: A novel multi-attention, multi-scale 3d deep network for coronary artery segmentation. Med. Image Anal. **85**, 102745 (2023)
6. Fayad, Z.A., et al.: Noninvasive in vivo human coronary artery lumen and wall imaging using black-blood magnetic resonance imaging. Circulation **102**(5), 506–510 (2000)
7. Gharleghi, R., Chen, N., Sowmya, A., Beier, S.: Towards automated coronary artery segmentation: a systematic review. Comput. Methods Programs Biomed. **225**, 107015 (2022)
8. Guo, C., Szemenyei, M., Yi, Y., Wang, W., Chen, B., Fan, C.: Sa-unet: spatial attention u-net for retinal vessel segmentation. In: 2020 25th International Conference on Pattern Recognition (ICPR), pp. 1236–1242. IEEE (2021)
9. Hodge, T., et al.: Effect of analyst segmentation variability on computed aaa stress distributions. In: International Conference on Medical Image Computing and Computer-Assisted Intervention, pp. 63–77. Springer (2021)

10. Jun, T.J., Kweon, J., Kim, Y.H., Kim, D.: T-net: nested encoder-decoder architecture for the main vessel segmentation in coronary angiography. Neural Netw. **128**, 216–233 (2020)
11. Mangiacapra, F., Bressi, E., Sticchi, A., Morisco, C., Barbato, E.: Fractional flow reserve (FFR) as a guide to treat coronary artery disease. Expert Rev. Cardiovasc. Ther. **16**(7), 465–477 (2018)
12. Prabha, D.S., Kumar, J.S.: Performance evaluation of image segmentation using objective methods. Indian J. Sci. Technol. **9**(8), 1–8 (2016)
13. Ronneberger, O., Fischer, P., Brox, T.: U-Net: convolutional networks for biomedical image segmentation. In: Navab, N., Hornegger, J., Wells, W.M., Frangi, A.F. (eds.) MICCAI 2015. LNCS, vol. 9351, pp. 234–241. Springer, Cham (2015). https://doi.org/10.1007/978-3-319-24574-4_28
14. Roth, G.A., et al.: Global burden of cardiovascular diseases and risk factors, 1990–2019: update from the GBD 2019 study. J. Am. Coll. Cardiol. **76**(25), 2982–3021 (2020)
15. Rygiel, P., Płuszka, P., Zieba, M., Konopczyński, T.: Centerlinepointnet++: a new point cloud based architecture for coronary artery pressure drop and vffr estimation. In: International Conference on Medical Image Computing and Computer-Assisted Intervention, pp. 781–790. Springer (2023)
16. Unser, M., Aldroubi, A., Eden, M., et al.: Fast b-spline transforms for continuous image representation and interpolation. IEEE Trans. Pattern Anal. Mach. Intell. **13**(3), 277–285 (1991)
17. Windecker, S., et al.: 2014 esc/eacts guidelines onmyocardial revascularization. J. Appl. physiol. **117**(3) (2014)
18. Zhou, Z., Rahman Siddiquee, M.M., Tajbakhsh, N., Liang, J.: Unet++: a nested u-net architecture for medical image segmentation. In: Deep Learning in Medical Image Analysis and Multimodal Learning for Clinical Decision Support: 4th International Workshop, DLMIA 2018, and 8th International Workshop, ML-CDS 2018, Held in Conjunction with MICCAI 2018, Granada, Spain, September 20, 2018, Proceedings 4. pp. 3–11. Springer (2018)

Conditional Graph Neural Network for Predicting Soft Tissue Deformation and Forces

Madina Kojanazarova(✉), Florentin Bieder, Robin Sandkühler, and Philippe C. Cattin

Department of Biomedical Engineering, University of Basel, Allschwil, Switzerland
M.Kojanazarova@unibas.ch

Abstract. Soft tissue simulation in virtual environments is becoming increasingly important for medical applications. However, the high deformability of soft tissue poses significant challenges. Existing methods rely on segmentation, meshing and estimation of stiffness properties of tissues. In addition, the integration of haptic feedback requires precise force estimation to enable a more immersive experience. We introduce a novel data-driven model, a *conditional graph neural network* (cGNN) to tackle this complexity. Our model takes surface points and the location of applied forces, and is specifically designed to predict the deformation of the points and the forces exerted on them. We trained our model on experimentally collected surface tracking data of a soft tissue phantom and used transfer learning to overcome the data scarcity by initially training it with mass-spring simulations and fine-tuning it with the experimental data. This approach improves the generalisation capability of the model and enables accurate predictions of tissue deformations and corresponding interaction forces. The results demonstrate that the model can predict deformations with a distance error of 0.35 ± 0.03 mm for deformations up to 30 mm and the force with an absolute error of 0.37 ± 0.05 N for forces up to 7.5 N. Our data-driven approach presents a promising solution to the intricate challenge of simulating soft tissues within virtual environments. Beyond its applicability in medical simulations, this approach holds the potential to benefit various fields where realistic soft tissue simulations are required.

Keywords: Soft tissue deformation · Graph neural networks · Haptic feedback

1 Introduction

Virtual reality (VR) has significantly impacted medical applications, including pre-operative planning, intraoperative navigation, training, and patient rehabilitation [2,22]. These immersive simulations enable medical professionals and trainees to interact with virtual anatomical structures, thereby improving understanding and increasing patient safety [6,21]. Combining haptic feedback systems

M. Kobielarz et al. (Eds.): MICCAI 2024, LNBE, pp. 62–73, 2025.
https://doi.org/10.1007/978-3-031-94128-3_7

with VR simulations enhances realism and interactivity, allowing users to manipulate virtual objects or tissues. Such realistic force feedback enables surgeons to perceive tissue resistance and other tactile sensations, leading to more effective training and improved surgical outcomes [5]. An example of such a visuo-haptic system was presented in [7], in which the patient-specific skull computed tomography (CT) volumes were visually and haptically rendered without resorting to mesh generation.

Accurately simulating soft tissue deformations is a significant challenge in VR-based anatomical simulations. Soft tissues exhibit complex mechanical behaviours, including elasticity, plasticity, and viscosity, making their simulation computationally demanding [10,15]. Traditional biomechanical methods such as the finite element method (FEM) and mass-spring models (MSMs) have been employed to simulate soft tissue deformations [25]. These methods often require time-consuming processes such as tissue segmentation from image volumes, mesh generation, and the estimation of tissue stiffness properties to simulate realistic behaviour [18,19,22]. Realistic haptic feedback simulations for deformable objects also rely on the outputs of the biomechanical methods [11,27]. Therefore, accurate and real-time interactive applications such as pre-operative planning are difficult to achieve.

Related Work. Research on modelling soft tissue deformation has mainly focused on biomechanical methods such as FEM, MSMs, and their variants [18]. While FEM simulations can achieve high accuracies, they are computationally expensive and too slow for real-time visuo-haptic simulations [18]. In contrast, MSMs offer faster calculations suitable for real-time applications but are less accurate than FEM simulations because their parameters do not directly correspond to the biomechanical properties of soft tissues [17]. Several solutions have been proposed to adapt MSMs to the properties of soft tissues [8,30].

Recently, data-driven approaches have addressed different limitations, e.g. time limitations, of biomechanical models for estimating soft tissue characteristics. Early methods used regression models to learn the biomechanical behaviour of liver and breast tissue [12], and to predict the deformation output of FEM simulations based on surface meshes of brain MR scans [26]. Other approaches have used image-based deep learning methods such as convolutional neural networks (CNNs) to predict deformations using FEM simulations of liver models for laparoscopic surgery simulation [13,14,20]. However, CNNs, which operate on regular grids, face challenges when applied to geometric data such as point clouds, graphs, and meshes, which are better suited for representing 3D objects with complex and irregular structures [1,3]. Graph neural networks (GNNs), in contrast, generalised convolutional operations on graph-like geometric data [24], and current GNNs use the message passing architecture. While the applications of GNNs have mainly focused on classification, segmentation, and generation tasks on conventional graph-like structures, their application to soft tissue simulation is a novel idea. The authors of [23] proposed PhysGNN, which uses GraphSAGE [9] and GraphConv [16] operators to predict the deformation of brain tissue under applied forces. They simulated deformations with FEM and used meshes as input graphs along with force and specific tissue properties, demonstrating that GNNs can accurately predict soft tissue deformations with low computational time using GPU acceleration.

Proposed Method. The paper proposes a proof of concept for a data-driven method, namely a *conditional Graph Neural Network* (cGNN), to simulate soft tissue deformations for VR applications with haptic feedback. Our method not only visually deforms the soft tissue at a given location but also estimates the interaction force, which is crucial for simulating haptic feedback in VR applications. By utilising a GNN, we leverage the power of point clouds and surface simulation to achieve accurate soft tissue modelling, overcoming the challenges associated with mesh generation, parameter estimation, and the preservation of detailed volumetric representations. To address the challenge of limited experimental data, we incorporated transfer learning. We initially pre-trained our model on simulated data generated using a mass-spring model (MSM) and fine-tuned it with experimentally collected surface tracking data of a soft tissue phantom. We aim to demonstrate that leveraging GNNs and transfer learning can make soft tissue simulation more accessible and efficient, especially with limited amounts of experimental data.

2 Methods

2.1 Experimental Data

We acquired the geometric and force data using a silicone phantom with size $100 \times 100 \times 32\,\text{mm}^3$ by mixing the two components of silicone rubber (Eurosil 4 A+B, Schouten SynTec, the Netherlands) with an oil softener S (Eurosil Softener, Schouten SynTec, the Netherlands) in a 1A:1B:2.5S ratio [29]. The deformations were captured using 25 pieces of 3 mm passive reflective markers (Fig. 1b) and a motion capture system consisting of five cameras (Qualisys AB, Sweden) at a sample rate of 500 Hz with a residual error of 0.18 ± 0.22 mm. The force feedback was captured simultaneously using a Nano17 force sensor with the calibration SI-12-0.12 (ATI Industrial Automation, USA). The sensor was attached to an indentation device with four markers to track the movement (Fig. 1a). The force measurement data was collected via a Speedgoat real-time target machine (Speedgoat, Switzerland) at a sample rate of 1000 Hz. The two measurements were synchronized by downsampling the force measurement to 500 Hz. We calculated the location of the tip of the indentation device using the Qualisys Track Manager 2021.2 and considered it the 26th marker on the surface. We collected the experimental dataset four times at five indentation locations each and then extracted static deformations at every 0.5 N for training and testing. This resulted in 244 static deformation states, with the max indentation force of 6.2 ± 1.4 N.

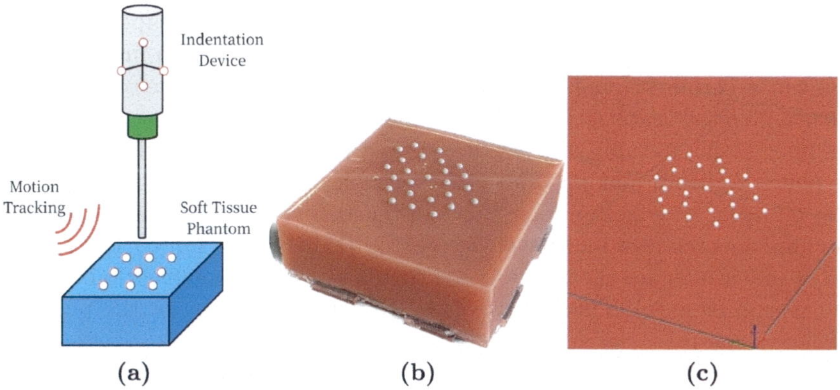

Fig. 1. Experimental measurement. **a** System setup, **b** silicone phantom with markers, and **c** measured surface points.

2.2 Simulated MSM Data

Mass-Springs and Parameter Optimisation. We simulated artificial soft tissue behaviour to create a larger dataset using mass-spring model (MSM) simulations. For this, we defined the soft tissue as a $100 \times 100\,\mathrm{mm}^3$ single-layer surface, simulated in a simple mass-spring model with linear between-mass springs and fixed-springs [4]. The surface consisted of 1024 points organised in a triangular 32×32 grid. With the masses set to 0.00016 kg, the damping factor 0.1, and the simulation time step of 0.0001 s, we optimised the spring parameters to match the behaviour of the measured deformations by minimising the distance error. The optimised parameters resulted in the between-mass spring coefficient 100 N/m and the fixed-spring coefficient 21 N/m.

Data Generation. To deform the surfaces and acquire a range from small to large deformations, we applied forces of up to $F_{max}= 7.5\,\mathrm{N}$ in $n_t= 15$ time steps and 11 directions. The directions include the surface normal and 10 randomly sampled directions up to 45° from the surface normal. The forces were applied at 100 randomly chosen locations on the surface, each location affecting $n_n = 1$ mass. The affected surface point received the following amount of force at the time t:

$$F_t = t \cdot \frac{1}{n_t \cdot n_n} F_{max}, \quad \text{for} \quad t \in \{1, \ldots, n_t\}. \tag{1}$$

We defined the system as stable when there were no significant changes in the velocity and the force of the masses ($v< 0.02\,\mathrm{m/s}$, $F< 0.02\,\mathrm{N}$). The data generation resulted in 16 500 static simulation examples.

2.3 Model

We developed a *conditional graph neural network* (cGNN) to predict the displacement of point clouds and corresponding forces. The cGNN model utilises the DynamicEdgeConv graph-convolutional operator [28] to capture the local geometric features of points while maintaining the permutation invariance.

Given a D-dimensional point cloud with N points, DynamicEdgeConv constructs a directed graph $\mathcal{G} = (\mathcal{V}, \mathcal{E})$ of a local point cloud structure as a k-nearest neighbour (k-NN) graph with vertices $\mathcal{V}$ and edges $\mathcal{E}$. Each node in the graph also points to itself, making $\mathcal{G}$ a self-loop graph. Edge features in the graph are collected using an edge function h, a non-linear function with learnable parameters implemented as a multilayer perceptron (MLP) [[28], Eq. 7]. The messages from edge features are then aggregated with a symmetric aggregation operation (sum, mean, or max). The output of the DynamicEdgeConv at the point x_i is therefore given by:

$$x_i' = \underset{j \in \mathcal{N}(i)}{\mathrm{AGGREGATE}}(h(x_i, x_j - x_i)), \tag{2}$$

where $\mathcal{N}(i) = \{j \in \mathcal{V} \mid (i, j) \in \mathcal{E}\}$ is the neighbourhood of point x_i.

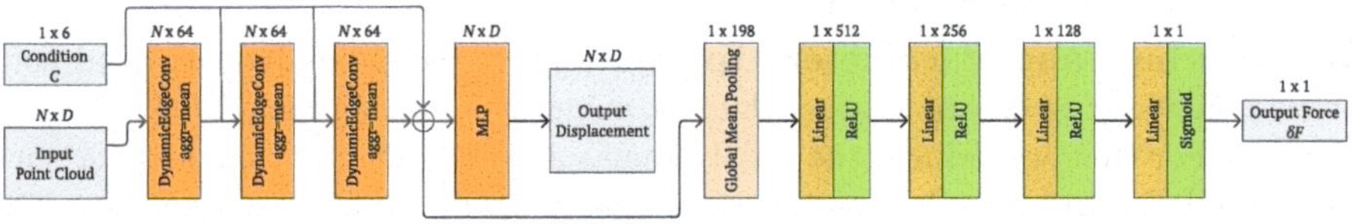

Fig. 2. The architecture of the *conditional graph neural network* (cGNN). $\oplus$ denotes the concatenation of the features

Our model (Fig. 2) takes a surface point cloud and extracts features through three DynamicEdgeConv operators. Concatenated outputs from DynamicEdgeConv layers and the specific conditions C of deformation are then passed through an MLP, which learns to predict the Euclidean displacement field with the same size as the input. The predicted displacement field can be applied to the input surface to acquire a deformed surface. Furthermore, the network extracts the global feature information from previously learned features, followed by four linear layers and activation functions to predict the change in force magnitude applied to the surface.

The input to the cGNN model are the N surface points with each $D = 3$ dimensions, given as Cartesian coordinates, and the deformation condition C. The condition $C \in \mathbb{R}^{2D}$ includes the point tip coordinates of the start c_s and end c_e of the applied tissue deformation defined as $C = [c_s,\ c_e]$. For an input point cloud $x \in \mathbb{R}^{N \times D}$, the output of the model f is given by $(\delta x, \delta F) = f(x, C)$, where $\delta x \in \mathbb{R}^{N \times D}$ is the predicted displacement field and $\delta F \in \mathbb{R}$ is the predicted change in force magnitude. The predicted displacement field is then applied to the input points to obtain the deformed surface $\hat{y} = x + \delta x$.

2.4 Training Setup

Loss Function. The learnable parameters of the model were trained by minimising the mean Euclidean distance $\mathcal{L}_d$ for deformed prediction, defined as:

$$\mathcal{L}_d = \frac{1}{N} \sum_{n=1}^{N} \|y_n - \hat{y}_n\|_2 \tag{3}$$

where N denotes the number of points, y_n is the position of the ground truth deformation, and $\hat{y}_n$ is the predicted deformation. The Mean square error (MSE) was used as a loss function $\mathcal{L}_f$ for the force predictions. The total loss was $\mathcal{L} = \alpha\mathcal{L}_d + \mathcal{L}_f$.

Hyperparameters. We trained the model for 200 epochs with a batch size of 32, and used the Adam optimiser with an initial learning rate of 10^{-4}. The loss weight for $\mathcal{L}_d$ was set to $\alpha = 88$. We used mean as the aggregation function in the DynamicEdgeConv layers in our model, and the number of neighbours k for the computation of the k-NN-graph was set to $k = 5$ as a trade-off between the model's prediction and time performance. Except for k, the hyperparameters were determined through a hyperparameter tuning using Optuna 3.5.0. The hyperparameters were not changed during the transfer learning.

Infrastructure. The MSM simulations were carried out on an AMD Epyc 7742 64-core CPU. The cGNN model was implemented using the open-source PyTorch-based machine learning library PyTorch Geometric 2.1.0. The models were trained and tested on an NVIDIA RTX 2080.

Table 1. Deformation depth features of the test sets, where δF is the force range in the dataset and δx is the displacement, total size referring to the size of the whole dataset

Dataset	Mode	Total size	δF [N]	δx_{max} [mm]	δx_{mean} [mm]
Experimental	Single-step	244	0.5	12.91	2.38 ± 2.97
	Multi-step	244	1 to 7.5	30.27	8.91 ± 7.96
MSM	Single-step	16 500	0.5	6.76	2.27 ± 1.13
	Multi-step	16 500	1 to 7.5	32.06	11.93 ± 6.66

Datasets and Transfer Learning. We divided our datasets into two modes, *Single-step* and *Multi-step*. The Single-step set includes all states of deformation from each time step t, with the ground truth as the deformation at $t+1$ ($\delta F = 0.5\,\text{N}$). The Multi-step set contains larger deformations and forces and includes 15 random states of t, where the ground truth was at least two time steps apart ($\delta F > 0.5$ N). Our model was trained on the combination of the two sets. These dataset modes were designed to train the model to handle various types of deformations and force increments. We split the data on the level of the deformation locations into training:validation:test sets with a ratio of 7:2:1 for the MSM dataset, and with a ratio of 12:3:5 for the experimental dataset. The ratio for the experimental dataset was chosen to ensure a broad representation in the test set, enhancing the diversity of examples from different recordings. In the experimental dataset, we performed a 5-fold cross-validation. Table 1 shows the test set features of the created datasets.

We trained cGNN with the simulated MSM dataset and compared the results by testing on the experimental dataset with and without transfer learning (TL). During TL, we fine-tuned the pre-trained model by continuing the training process with experimental data. In an ablation study, we tested the effect of the conditional encoding of our model by eliminating the deformation condition C as an input, referred to as *noCond.* To test the model's predictive consistency with respect to the number of training points used, we trained the model with augmented MSM data (aug MSM), by inputting 10 % of the randomly selected surface points for each sample during the training, and further testing it with the total amount of points.

Table 2. Performance of cGNN on the mass-spring method dataset (MSM) and the experimental dataset (Exp) with and without transfer learning (TL). The model noCond indicates that cGNN was trained without conditional encoding

Model/ Train set	Test set	Force MSE [N^2]	Force Absolute Error [N]	Mean $\mathcal{L}_d$ [mm]	Max $\mathcal{L}_d$ [mm]
noCond/MSM	MSM	2.20	1.19 ± 0.88	0.56 ± 0.21	5.43 ± 5.64
cGNN/MSM	MSM	$4 \cdot 10^{-3}$	0.03 ± 0.06	0.43 ± 0.11	2.22 ± 2.11
cGNN/aug MSM	MSM	$5 \cdot 10^{-3}$	0.03 ± 0.06	0.46 ± 0.12	2.35 ± 2.16
cGNN/Exp	Exp	0.48	0.48 ± 0.08	0.47 ± 0.03	3.06 ± 0.55
cGNN/MSM (TL)	Exp	0.39	0.41 ± 0.02	0.36 ± 0.02	**2.03 ± 0.29**
cGNN/aug MSM (TL)	Exp	**0.35**	**0.37 ± 0.05**	**0.35 ± 0.03**	2.10 ± 0.35

3 Results and Discussion

Performance of cGNN. The results in Table 2 show that our model predicted the displacement with a 0.43 ± 0.11 mm Euclidean error and an absolute force error of 0.03 ± 0.06 N for the combined MSM dataset. The training of our model without the conditional encoding (noCond) significantly decreased the performance for both displacement and force predictions, showing the importance of conditioning when generalising predictions. In contrast, training cGNN with the augmented MSM dataset (aug MSM), i.e. using 10 % of the randomly selected surface points for each sample, did not noticeably affect the performance when tested on the complete set of points.

The implementation of TL for the experimental dataset significantly improved the deformation prediction compared to training without TL, resulting in a reduction of 0.11 mm (23 %) of the mean displacement error and a reduction of 1.03 mm (34 %) of the maximal displacement error with TL compared to without TL. As for the force prediction, the performance was similar, resulting in lower force prediction errors with TL. Figure 3 shows the deformation results where the predicted deformation field was applied to the input points. Creating an MSM simulation took 2.9321 ± 0.5451 s, while predicting it with the proposed cGNN took 0.0947 ± 0.0003 s. For each experimental data prediction, the cGNN took only 0.0029 ± 0.0001 s.

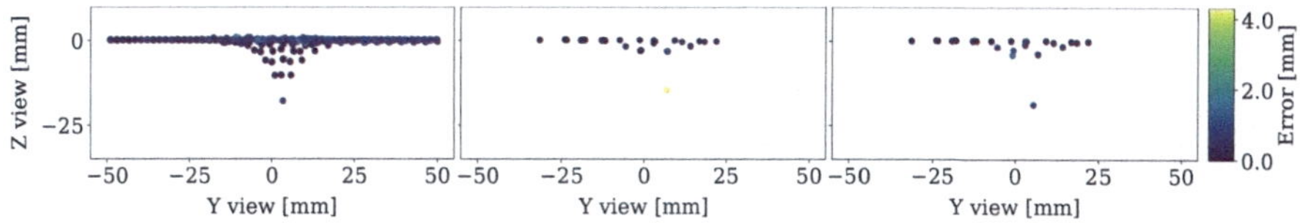

(a) MSM δF_{pred}=1.51 N (b) Exp δF_{pred}=1.57 N (c) Exp (TL) δF_{pred}=1.24 N

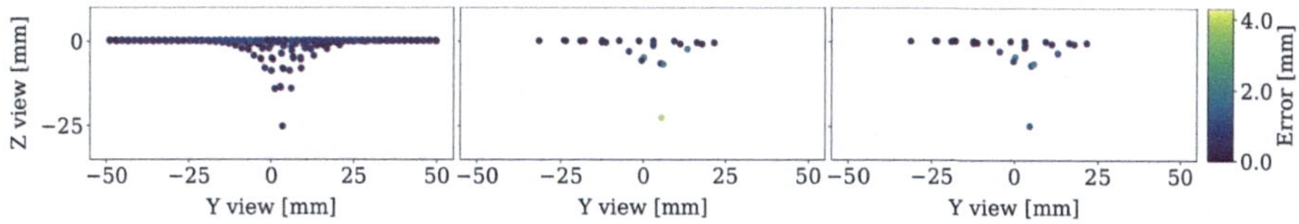

(d) MSM δF_{pred}=4.49 N (e) Exp δF_{pred}=4.65 N (f) Exp (TL) δF_{pred}=4.47 N

Fig. 3. Prediction results. First row: ground truth force δF= 1.50 N, second row: ground truth force δF= 4.50 N for **a, d** mass-spring simulation, **b, e** experimental data without transfer learning and **c, f** experimental data with transfer learning.

Data Scarcity. Due to the limited experimental data, we also trained cGNN with varying amounts of training samples, both with and without TL. Each training sample included all static data from one indentation location. As shown in Fig. 4, using TL resulted in lower errors in deformation prediction, with optimal performance achieved using only 7–8 training samples. Additionally, TL with the model pre-trained on augmented MSM data demonstrated consistent performance for displacement predictions and significantly improved the force predictions.

Limitations and Implications. The main limitation of this study is that we simplified our soft tissue model to a single surface layer of homogeneous tissues to avoid meshing. This simplification was chosen to demonstrate the feasibility of our approach and provide a clear proof of concept. Further analysis is needed

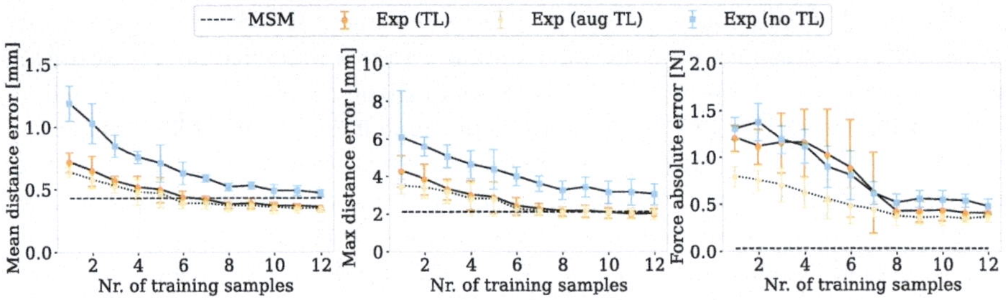

Fig. 4. Mean distance error, max distance error and force absolute error of experimental data with transfer learning (Exp TL), without transfer learning (Exp no TL) and augmented transfer learning (Exp aug TL) when trained on different amounts of training samples compared to MSM simulations.

to investigate the behaviour of inhomogeneous tissue with depth information, possibly using voxel intensities from CT volumes.

Additionally, the high computational time of our network when predicting MSM simulations with 1024 points is a significant limitation. This is primarily due to the DynamicEdgeConv layers, which involve expensive calculations of dynamic edge features in each layer. Despite this, we chose this graph-convolutional layer to develop a model capable of simulating deformations based solely on surface points, thereby eliminating the need for meshing the object. Notably, while the computation time for the simulated data was high, it was reasonable for the experimental data due to the smaller number of points. This reflects a realistic scenario where experimental data often involves fewer points, indicating that computation time is not a critical issue in practical applications.

The use of transfer learning on limited experimental data highlights an important implication of our work. We demonstrated that our model, initially trained on simulated data, could be effectively fine-tuned with a small set of experimental data. This approach mitigates the issue of data scarcity, showing that even with limited data, the model can be adapted to new datasets. Our "data scarcity" test revealed that the model performs well with as few as 7–8 examples, suggesting that future applications can easily adopt this approach when additional data becomes available.

It is important to note that our work and results cannot be directly compared with other studies due to the different approaches and data used. Other methods, such as [23], used larger meshes and FEM simulations, where computation time was not an issue, but they required the simulation of biomechanical properties. Their model requires an input of tissue properties and the applied force to predict deformations. In contrast, our method predicts deformations and forces without requiring detailed tissue information, simplifying the process and avoiding the computational complexity associated with biomechanical property estimation. Our study demonstrates a versatile approach to soft tissue surface deformation that can predict the force exerted on the tissue without requiring additional information about the tissue properties.

4 Conclusions

This study aimed to propose a novel graph-based and data-driven simulation method, cGNN, to generalise soft tissue deformation for VR applications that use haptic force feedback. We demonstrated the ability of cGNN to successfully deform synthetic and real soft tissue surfaces using a transfer learning approach. Our model also predicts the force applied to them by conditioning them to the start and end location of the exerted force. By showing that our method can accurately predict deformations using a simple single-layer model, we provide a foundation for future developments in more complex tissue simulations. This proof of concept highlights the potential of our approach to enhance the realism and effectiveness of VR-based medical training and planning tools.

Acknowledgments. The authors would like to acknowledge the support from Felix Erb and the BIROMED-Lab (Department of Biomedical Engineering, University of Basel) for the assistance with the experimental data measurement.

References

1. Ahmed, E., et al.: A survey on deep learning advances on different 3D data representations. arXiv:1808.01462 (2018)
2. Bin, S., Masood, S., Jung, Y.: Chapter twenty - Virtual and augmented reality in medicine. In: D.D. Feng (ed.) Biomedical Information Technology (Second Edition), Biomedical Engineering, 2nd edn., pp. 673–686. Academic Press (2020)
3. Bronstein, M.M., Bruna, J., LeCun, Y., Szlam, A., Vandergheynst, P.: Geometric deep learning: Going beyond Euclidean data. IEEE Signal Process. Mag. **34**(4), 18–42 (2017)
4. Chen, F., Gu, L., Huang, P., Zhang, J., Xu, J.: Soft tissue modeling using nonlinear mass spring and simplified medial representation. In: 2007 29th Annual International Conference of the IEEE Engineering in Medicine and Biology Society, pp. 5083–5086 (2007)
5. Coles, T.R., Meglan, D., John, N.W.: The role of haptics in medical training simulators: a survey of the state of the art. IEEE Trans. Haptics **4**(1), 51–66 (2011)
6. Dakson, A., Hong, M., Clarke, D.B.: Virtual Reality Surgical Simulation: Implications for Resection of Intracranial Gliomas, vol. 30, pp. 106–116. Karger Publishers, Basel (2017)
7. Faludi, B., Zoller, E.I., Gerig, N., Zam, A., Rauter, G., Cattin, P.C.: Direct visual and haptic volume rendering of medical data sets for an immersive exploration in virtual reality. In: Shen, D., et al. (eds.) MICCAI 2019. LNCS, vol. 11768, pp. 29–37. Springer, Cham (2019). https://doi.org/10.1007/978-3-030-32254-0_4
8. Golec, K.: Hybrid 3D Mass Spring System for Soft Tissue Simulation. Université de Lyon, Thesis (2018)
9. Hamilton, W.L., Ying, R., Leskovec, J.: Inductive representation learning on large graphs. In: Proceedings of the 31st International Conference on Neural Information Processing Systems, NIPS 2017, pp. 1025–1035 (2017)
10. Holzapfel, G.: Biomechanics of Soft Tissue, Volume III, Multiphysics Behaviors, Chapter 10, Composite Media, pp. 1049–1063. Academic Press, United States (2001)

11. Kim, S.Y., Park, J., Kwon, D.S.: Area-contact haptic simulation. In: Ayache, N., Delingette, H. (eds.) Surgery Simulation and Soft Tissue Modeling, pp. 108–120 (2003)
12. Martín-Guerrero, J.D., et al.: Machine learning for modeling the biomechanical behavior of human soft tissue. In: 2016 IEEE 16th International Conference on Data Mining Workshops (ICDMW), pp. 247–253 (2016)
13. Mendizabal, A., Márquez-Neila, P., Cotin, S.: Simulation of hyperelastic materials in real-time using deep learning. Med. Image Anal. **59**, 101,569 (2020)
14. Mendizabal, A., Tagliabue, E., Hoellinger, T., Brunet, J.-N., Nikolaev, S., Cotin, S.: Data-driven simulation for augmented surgery. In: Developments and Novel Approaches in Biomechanics and Metamaterials. ASM, vol. 132, pp. 71–96. Springer, Cham (2020). https://doi.org/10.1007/978-3-030-50464-9_5
15. Miller, K.: Biomechanics of soft tissues. Medical science monitor. Int. Med. J. Exp. Clin. Res. **6**(1), 158–67 (2000)
16. Morris, C., et al.: Weisfeiler and leman go neural: higher-order graph neural networks. AAAI **33**(01), 4602–4609 (2019)
17. Nealen, A., Müller, M., Keiser, R., Boxerman, E., Carlson, M.: Physically based deformable models in computer graphics. Comput. Graph. Forum **25**(4), 809–836 (2006)
18. Nguyen, T.N., Ho Ba Tho, M.C., Dao, T.T.: A systematic review of Real-Time medical simulations with soft-tissue deformation: computational approaches, interaction devices, system architectures, and clinical validations. Appl. Bionics Biomech. **2020** (2020)
19. Payan, Y.: Soft tissue finite element modeling and calibration of the material properties in the context of computer-assisted medical interventions. In: Avril, S., Evans, S. (eds.) Material Parameter Identification and Inverse Problems in Soft Tissue Biomechanics. CICMS, vol. 573, pp. 133–144. Springer, Cham (2017). https://doi.org/10.1007/978-3-319-45071-1_6
20. Pfeiffer, M., Riediger, C., Weitz, J., Speidel, S.: Learning soft tissue behavior of organs for surgical navigation with convolutional neural networks. Int. J. Comput. Assist. Radiol. Surg. **14**(7), 1147–1155 (2019). https://doi.org/10.1007/s11548-019-01965-7
21. Pottle, J.: Virtual reality and the transformation of medical education. Future Healthc J **6**(3), 181–185 (2019)
22. Preim, B., Botha, C.: Chapter e21 - Computer-assisted medical education. In: Preim, B., Botha, C. (eds.) Visual Computing for Medicine (Second Edition), 2nd edn., pp. e101–e169. Morgan Kaufmann, Boston (2014)
23. Salehi, Y., Giannacopoulos, D.: PhysGNN: a physics–driven graph neural network based model for predicting soft tissue deformation in image–guided neurosurgery. In: Oh, A.H., Agarwal, A., Belgrave, D., Cho, K. (eds.) Advances in Neural Information Processing Systems (2022)
24. Scarselli, F., Gori, M., Tsoi, A.C., Hagenbuchner, M., Monfardini, G.: The graph neural network model. IEEE Trans. Neural Netw. **20**(1), 61–80 (2009)
25. Schill, M.A.: Biomechanical soft tissue modeling - Techniques, implementation and application. Ph.D. thesis, University of Mannheim (2001)
26. Tonutti, M., Gras, G., Yang, G.Z.: A machine learning approach for real-time modelling of tissue deformation in image-guided neurosurgery. Artif. Intell. Med. **80**, 39–47 (2017)
27. Vaughan, N., Dubey, V., Wee, M., Isaacs, R.: Haptic feedback from human tissues of various stiffness and homogeneity. Adv. Robot. Res. **1**, 215–237 (2014)

28. Wang, Y., Sun, Y., Liu, Z., Sarma, S.E., Bronstein, M.M., Solomon, J.M.: Dynamic graph CNN for learning on point clouds. ACM Trans. Graph. **38**(5), 1–12 (2019)
29. Yushchenko, M., Sarracanie, M., Amann, M., Sinkus, R., Wuerfel, J., Salameh, N.: Elastography validity criteria definition using numerical simulations and MR acquisitions on a low-cost structured phantom. Front. Phys. **9** (2021)
30. Zerbato, D., Galvan, S., Fiorini, P.: Calibration of mass spring models for organ simulations. In: 2007 IEEE/RSJ International Conference on Intelligent Robots and Systems, pp. 370–375 (2007)

An Efficient Pipeline to Compute Patient-Specific Cerebral Aneurysm Wall Tension

Mostafa Jamshidian[1(✉)], Benjamin F. Zwick[1], Arosha S. Dissanayake[2], Adam Wittek[1], Timothy J. Phillips[3], Stephen Honeybul[2], Graeme J. Hankey[4,5], and Karol Miller[1]

[1] Intelligent Systems for Medicine Laboratory, The University of Western Australia, Perth, WA, Australia
mostafa.jamshidian@uwa.edu.au
[2] Department of Neurosurgery, Sir Charles Gairdner Hospital, Perth, WA, Australia
[3] Neurological Intervention and Imaging Service of Western Australia, Sir Charles Gairdner Hospital, Perth, WA, Australia
[4] Centre for Neuromuscular and Neurological Diseases, Medical School, The University of Western Australia, Perth, WA, Australia
[5] Perron Institute for Neurological and Translational Science, Perth, WA, Australia

Abstract. Cerebral aneurysm rupture, leading to subarachnoid hemorrhage with a high mortality rate, disproportionately affects younger populations, resulting in a significant loss of productive life years. A significant proportion of these deaths is due to aneurysmal re-bleeding within the first three days following the initial bleed, prior to treatment. While early aneurysm treatment is recommended, there is no consensus on the ideal timing, and emergency treatment offers only an incremental benefit at a significant cost. Although various multivariable prediction models have been proposed to provide personalized risk assessments, no validated patient-specific predictor is available to rationalize emergency treatment. Furthermore, no model has yet incorporated emerging computational biomechanics-based biomarkers such as wall tension. In this paper, we proposed and validated an efficient semi-automatic pipeline to compute patient-specific cerebral aneurysm wall tension as a potential biomarker for the likelihood of re-bleeding. Our pipeline uses the patient's computed tomography angiography (CTA) image obtained at the time of subarachnoid hemorrhage diagnosis to create a patient-specific biomechanical model of the cerebral aneurysm using the finite element method. A distinctive feature of our approach is the straightforward model creation and wall tension computation using shell finite elements, without requiring patient-specific material properties or aneurysm wall thickness. Our non-invasive, patient-specific method for cerebral aneurysm wall tension can potentially provide individualized risk prediction and enhance clinical decision-making.

Keywords: Cerebral Aneurysm · Biomechanics · Patient-specific Analysis

M. Kobielarz et al. (Eds.): MICCAI 2024, LNBE, pp. 74–87, 2025.
https://doi.org/10.1007/978-3-031-94128-3_8

1 Introduction

Aneurysmal subarachnoid hemorrhage (aSAH), often due to the rupture of a saccular cerebral aneurysm [1], disproportionately affects a younger demographic [2], resulting in the loss of productive life years, which makes its impact nearly equivalent to that of the more common ischemic stroke [3]. One quarter of aSAH patients will die prior to reaching hospital [4], and up to one in six hospitalized patients will die prior to treatment within the first 72 h, most often due to re-bleeding [5].

Whilst aneurysm treatment as soon as feasible is recommended, there is no consensus regarding the ideal timing due to the complex interplay between time-to-treatment and other risk factors, such as the degree of conscious state impairment and the thickness and distribution of the bleed. Routine emergency aneurysm treatment results in only a small incremental benefit but at a significant cost [6]. To rationalize emergency treatment, various multivariable prediction models have been proposed to provide personalized risk assessments, but they have not been systematically evaluated or externally validated. Additionally, these models cannot provide predictions for approximately 20% of patients with multiple causative aneurysms [7]. To date, no model has integrated emerging biomarkers such as aneurysm wall tension calculated by computational biomechanics [8] and early brain injury (EBI) markers [9, 10], although stress analysis of cerebral aneurysms has been researched previously [11–16].

In this study, we present an efficient semi-automatic software pipeline to compute patient-specific cerebral aneurysm wall tension as a possible biomarker for the likelihood of re-bleeding. In our pipeline, we use computed tomography angiography (CTA) images to create patient-specific biomechanical models of the cerebral aneurysms using finite element (FE) method, without requiring patient-specific material properties and aneurysm wall thickness. Specific to our approach is the implementation and verification of a straightforward approach for computing patient-specific wall tension in cerebral aneurysms using shell elements.

The remainder of the paper is organized as follows: In Sect. 2, we present the methods for wall tension computation, surface model creation, and biomechanical model creation using shell elements. We also describe the verification of the shell FE model against a solid FE model and the verification of our wall tension computation method. In Sect. 3, we present the results for the mesh convergence analysis, verification of the shell FE model, and verification of our approach for wall tension computation. Finally, Sect. 4 is devoted to the conclusions and discussions.

2 Methods

Based on mechanics, an artery ruptures when the local wall stress exceeds the local wall strength. Therefore, cerebral aneurysm wall stress, determined reliably and quickly through non-invasive approaches, can aid in patient-specific disease management. To compute cerebral aneurysm wall tension, we followed the approach implemented in the freely available BioPARR (Biomechanics-based Prediction of Aneurysm Rupture Risk) software package [17, 18]. Originally developed for computing wall stress in abdominal aortic aneurysms, BioPARR uses geometry from CTA images and mean arterial pressure as the applied load.

BioPARR stress recovery method, using linear FE analysis, is based on the observation that, as patient images demonstrate a deformed (due to blood pressure loading) artery geometry, the internal forces on the blood vessel wall that balance the applied pressure can be calculated based on statics principles [17, 19]. Accordingly, if the biomechanics simulations are arranged so that the aneurysm geometry remains unchanged when loaded by blood pressure, the stress field in the aneurysm wall balances the pressure [19]. Moreover, this stress field is shown to be independent of the material properties of artery tissue if the material is homogeneous in the current geometric configuration [20–24], although some studies have shown that this is also true for heterogeneous tissues [25].

In BioPARR, residual stresses are incorporated using Fung's Uniform Stress Hypothesis [26, 27], which involves averaging the stress across the artery wall thickness. This approach simplifies the computation of residual stresses, making it a straightforward post-processing step in aneurysm wall stress analysis [19]. We previously verified that, for a given material model and properties, the residual stress fields obtained using the BioPARR method are very close to those produced by the more complex non-linear iterative stress analysis method [26, 28].

2.1 Wall Tension

Joldes et al. [26] proposed and validated a simple approach to compute arterial wall stress, incorporating residual stresses according to Fung's Uniform Stress Hypothesis [27]. The maximum principal wall stress $\overline{\sigma}$ based on the Uniform Stress Hypothesis is given by [26]:

$$\overline{\sigma} = \frac{1}{t} \int_0^t \sigma(\xi) d\xi, \tag{1}$$

where $\sigma(\xi)$ for $0 \leq \xi \leq t$ is the maximum principal stress (MPS) across the wall thickness t. Miller et al. [29] showed that maximum principal wall stress is proportional to the wall thickness. Therefore, inaccuracies in the difficult-to-measure wall thickness directly translate into errors in wall stress computation [30].

Considering the measurement difficulties, high uncertainties, and variations in patient-specific wall thickness of cerebral aneurysms, we used the maximum principal wall tension (MPWT) or simply wall tension, which is independent of wall thickness and is given by:

$$\overline{T} = \int_0^t \sigma(\xi) d\xi, \tag{2}$$

with units of force per length.

In FE analysis, MPWT is computed by numerical integration of MPS through thickness. Using an unstructured tetrahedral solid FE mesh, [26] computed the integral in Eq. (2) as a sum of piecewise integrals evaluated on several smaller sub-intervals of the wall thickness, using a two-point Gauss rule on each sub-interval. However, using a shell FE mesh, we have shown and verified in Sect. 3 that MPWT directly results from the outputs of the shell elements, with no numerical integration.

2.2 Image Segmentation and Surface Model

We used the 3D Slicer image computing platform [31, 32] to create the surface triangulation model of the cerebral aneurysm. We cropped a region of interest, encompassing the cerebral arteries with aneurysm, from Patient 1's fine-slice (voxel size 0.4 $\sim$ 0.5 mm) CTA image obtained at the time of aSAH diagnosis, as shown in Fig. 1a. We semi-automatically segmented the geometry of cerebral arteries with aneurysm and created the surface model from the segmentation, as shown in Fig. 1b. Using the Vascular Modeling Toolkit (VMTK) [33] extension in 3D Slicer, we extracted the centerlines of vessels and clipped vessels to create the surface model of cerebral aneurysm with clipped connecting vessels, as shown in Fig. 1c. We saved this final surface triangulation model as an STL file and used it as the starting point for generating the FE mesh.

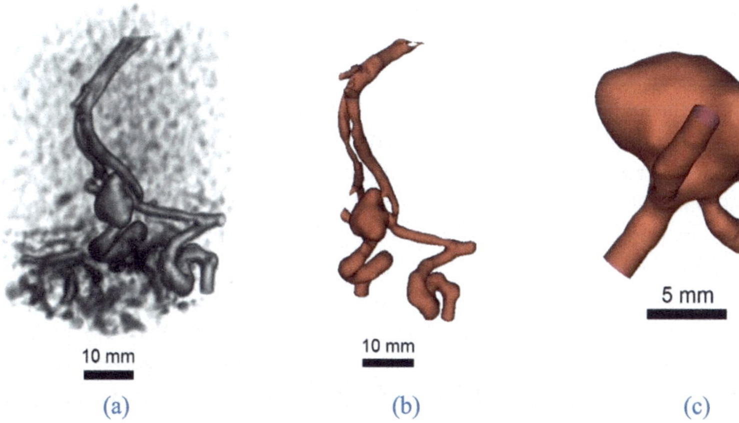

Fig. 1. Patient 1's cerebral aneurysm: (a) Volume rendering of the cropped 3D CTA image containing the cerebral arteries with aneurysm, (b) surface model of the cerebral arteries with aneurysm, (c) surface model of the cerebral aneurysm with clipped connecting vessels.

2.3 Biomechanical Model

Using BioPARR methods, we computed the patient-specific wall tension of cerebral aneurysm through a direct linear analysis with no patient-specific material property data [19]. Following [11–13], we used shell elements in our FE analysis. Shell elements offer a significant computational advantage over solid elements for the analysis of thin-walled structures. By employing dimensional reduction, treating thickness as a parameter, and making simplified kinematic assumptions, shell elements require fewer elements and nodes. Aside from the simplicity of FE model creation, another major advantage of using shell elements for our application is the ease of computing wall tension from the output results of shell elements, as will be described later in Sect. 2.5.

To ensure the simplified kinematic assumptions do not affect the stress recovery results, we verified the stress recovery results of the shell FE analysis against those of a solid FE analysis for the same Patient 1's aneurysm geometry. Providing a full three-dimensional representation of the structure, solid elements can generally provide more accurate stress analysis for complex geometries and boundary conditions. We used the STL surface model of the aneurysm in Fig. 1c as the lumen boundary and created the aneurysm wall model by outwardly offsetting this surface. For the FE analysis, we used the Abaqus FE software [34].

Shell FE Mesh. We re-meshed the surface model of the aneurysm in Fig. 1c, using VMTK to create a high-quality triangular shell mesh with desired element size. For mesh convergence analysis, we created shell meshes using four different element sizes of 0.15, 0.20, 0.3, and 0.6 mm, as shown in Fig. 2. We imported the shell mesh into Abaqus via the built-in STL Import plug-in. We used the Abaqus STRI65 quadratic shell element that is a 6-node triangular thin shell, with five degrees of freedom per node. The five degrees of freedom per node for the Abaqus STRI65 shell element are translations in the local x (in-plane), y (in-plane), and z (out-of-plane) directions, and rotations about the local x and y axes.

The STRI65 element has three integration points in the element plane. To obtain stress variation through the shell thickness as output, we instructed Abaqus to perform section integration during analysis. Abaqus conducted numerical integration at a user-defined odd number of integration points through thickness to calculate the stresses independently at each integration point. These through-thickness integration points are located at the positions of the in-plane integration points and are evenly distributed along the normal to the shell element plane. With an odd number of integration points, the first and last integration points are located at the innermost and outermost positions along the thickness direction, and the middle integration point is at the middle of the thickness.

Seshaiyer et al. [14]. Reported cerebral aneurysm wall thicknesses between 44 μm and 128 μm, with an average wall thickness of 86 μm. Since wall tension is independent of wall thickness, we used a uniform shell thickness parameter of 0.086 mm [11, 12], with outwards thickness offsetting from the element plane for all elements. The outwards thickness offsetting implies that the shell mesh, shown in Fig. 2, represents the internal surface of the thin-walled vessel.

Solid FE Mesh. We used Gmsh [35, 36] to generate a high-quality solid mesh of wedge elements, with specified in-plane element size and desired number of element layers through the thickness, based on the STL surface model of aneurysm in Fig. 1c. We instructed Gmsh to thicken the solid mesh by outwards offsetting so that the STL surface model, serving as the basis for solid mesh generation, overlapped with the internal surface of the solid mesh. For mesh convergence analysis, we created solid meshes with four different refinements including, the in-plane element sizes of 0.15 mm and three element layer through thickness, the in-plane element sizes of 0.20 mm and two element layer through thickness, the in-plane element sizes of 0.30 mm and two element layer through thickness, and the in-plane element sizes of 0.60 mm and one element layer through thickness. Consistent with the shell mesh, the thickness of the solid mesh for all four meshes was 0.086 mm [11, 12]. It should be noted that, as shown in Fig. 2, the solid mesh explicitly models the thickness, whereas the thickness in the shell mesh is a parameter. The solid mesh was exported from Gmsh as an Abaqus input file and imported into Abaqus as an orphan solid mesh. We used the Abaqus C3D15 continuum element, that is a 15-node quadratic wedge (triangular prism).

We used a linear elastic material model with a Poisson's ratio of 0.49 for the vessel wall tissue [11, 37]. We used a large elastic modulus of 100 GPa in our linear FE analysis for stress recovery.

We fixed the clipped ends of the connecting vessels and applied a uniform pressure of 13.332 kPa (100 mmHg) on the inner surface of the aneurysm [11]. We solved the FE model of Patient 1's cerebral aneurysm using Abaqus/Standard solver [34].

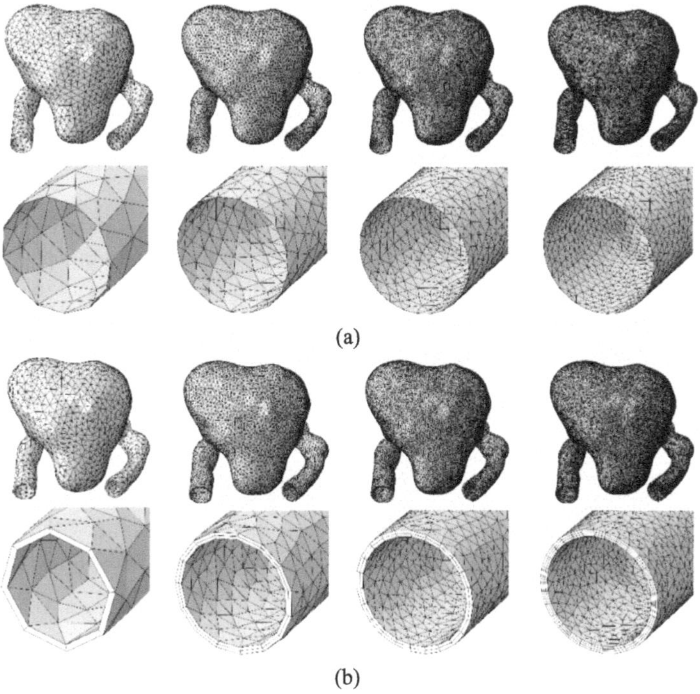

Fig. 2. (a) Shell and (b) solid finite element meshes used for mesh convergence analysis. Mesh refinement increases from left to right, with coarser meshes on the left and finer meshes on the right.

2.4 Method Verification

For verification of mesh refinement via mesh convergence analysis, verification of shell mesh results against the solid mesh results, and verification of MPWT as a direct output of the shell FE analysis, we used stress percentile plots to compare two sets of stress results on the same domain (aneurysm wall) but represented on different sets of discrete points [38–40]. In the stress percentile plots, the horizontal axis indicates the percentile rank, and the vertical axis shows the corresponding percentile value.

Verification of Mesh Refinement. We performed the mesh convergence analysis for solid and shell meshes separately, each with four different mesh refinements, as shown in Fig. 2. For verification, we compared the percentile plots of the MPS instead of the MPWT because we did not implement the numerical computation of MPWT for the solid mesh.

Verification of Shell Mesh. We verified the shell mesh FE analysis by comparing the percentile plots of MPS on the internal surface of the aneurysm between the shell and solid meshes, as they both originate from and share the same internal surface, as shown in Fig. 1c.

For the shell mesh, we created the MPS percentile plot using the MPS values at the innermost through-thickness integration points, as these points represent the internal surface of the aneurysm. However, in the solid mesh, the integration points lie within the thickness. We instructed Abaqus to report the MPS values on the internal surface of the solid mesh by interpolation using element shape functions. For the solid mesh, we created the MPS percentile plot using the MPS values at the interior FE nodes representing the internal surface.

It should be noted that in Abaqus, shell element outputs include both the maximum principal stress and the maximum in-plane principal stress. In this paper, for shell elements, MPS stands for the maximum in-plane principal stress output in Abaqus.

Verification of Wall Tension: For the numerical computation of MPWT for shell elements using Eq. (2), we instructed Abaqus to output the stresses at all through-thickness integration points. Using Eq. (2), we computed MPWT via numerical integration as follows:

$$\overline{T} = \frac{t}{n}\sum_{i=1}^{n} \sigma(i), \tag{3}$$

where $\sigma(i)$ is the MPS at the i th through-thickness integration point, with $i = 1, \ldots, n$. For precise computation of MPWT, we used $n = 15$ through-thickness integration points.

It should be noted that the shell section forces, as a direct output from Abaqus, cannot be used to compute MPWT because these forces are provided in a local coordinate system that does not align with the MPS direction. According to Eq. (2), MPWT is the shell section force in the MPS direction.

Instead of numerical integration using Eq. (3), we have shown and verified in Sect. 3 that, in a linear analysis using a shell mesh, MPWT equals the MPS at the mid-surface through the thickness multiplied by the shell thickness. The MPWT field output was created in Abaqus by multiplying the shell thickness by the MPS field output at the mid through-thickness integration points, which is a direct output from Abaqus.

To verify this technique, we compared the percentile plots of MPWT computed via numerical integration using Eq. (3) with MPWT computed as MPS at mid through-thickness integration points multiplied by the shell thickness.

3 Results

Figures 3 and 4 show the mesh convergence analysis results for solid and shell meshes, respectively, using the FE meshes shown in Fig. 2. The MPS on the internal surface of the cerebral aneurysm was used as the variable of interest in the mesh convergence analysis. In each figure, the MPS percentile plots illustrate convergence across various mesh refinements. From left to right, the MPS contour plots correspond to increasingly finer meshes. Figures 3 and 4 demonstrate that the solution converges at an FE size of 0.3 mm for both shell and solid FE meshes. To account for variations in aneurysm geometry, we used an FE size of 0.2 mm as the sufficient mesh refinement for both shell and solid FE models. The Shell mesh with FE size of 0.2 mm consists of 74292 nodes and 37078 elements of type STRI65. The solid mesh with FE size of 0.2 mm consists of 212565 nodes and 60601 elements of type C3D15. Additionally, a separate convergence analysis on the number of through-thickness integration points determined that five integration points through the shell thickness are sufficient. These results are not presented here for brevity.

The simulation times for shell and solid FE models are 30 s and 70 s, respectively, on a laptop workstation with a 12th Gen Intel® Core™ i9-12900H 2.50 GHz processor and 64.0 GB of RAM. Although the difference in simulation time might be negligible for practical applications, the shell FE model is advantageous due to its straightforwardness in modeling and wall tension extraction.

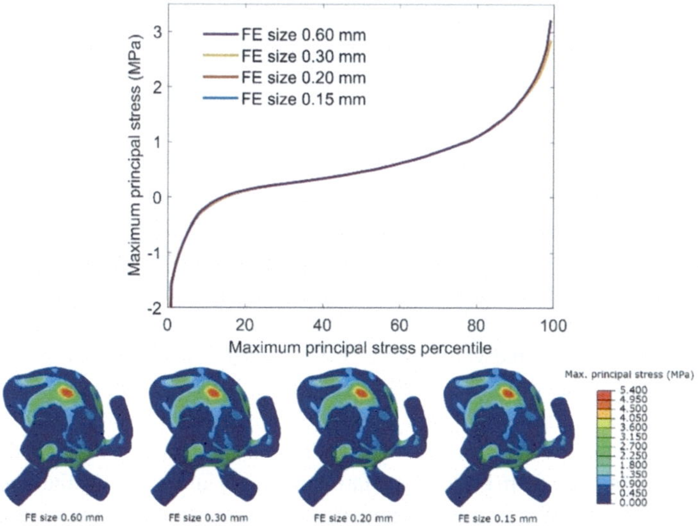

Fig. 3. Mesh convergence analysis for the shell finite element mesh using the maximum principal stress (MPS) on the internal surface of the cerebral aneurysm as the variable of interest. The MPS percentile plots show convergence across various mesh refinements. The MPS contour plots, from left to right, correspond to coarser to finer meshes.

Figure 5 shows the verification of the shell FE model solution against the solid FE model solution by comparing their results for the MPS on the internal surface of the cerebral aneurysm shown in Fig. 1c. In addition to the MPS contour plots, Fig. 5 includes percentile plots of MPS, providing a detailed comparison between the two meshes. Figure 5 demonstrates that the shell mesh produces almost identical MPS results to the solid mesh, except for MPS percentile ranks below the 15th.

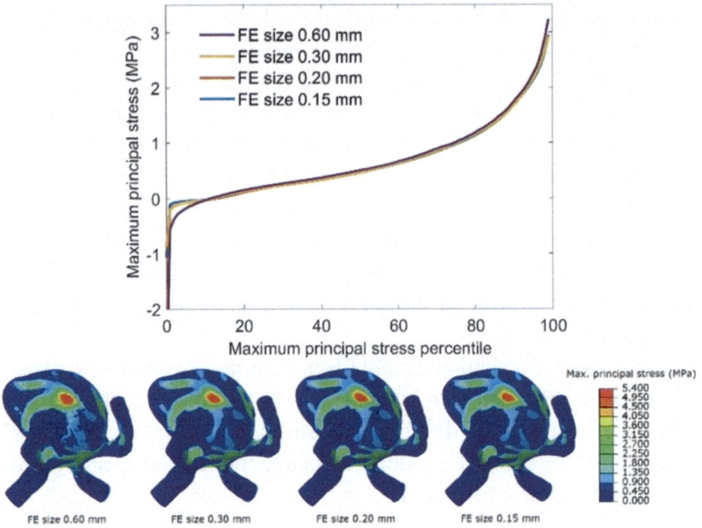

Fig. 4. Mesh convergence analysis for the solid finite element mesh using the maximum principal stress (MPS) on the internal surface of the cerebral aneurysm as the variable of interest. The MPS percentile plots show convergence across various mesh refinements. The MPS contour plots, from left to right, correspond to coarser to finer meshes.

Since the aim of our stress analysis is to assess rupture risk by comparing the maximum local wall stress against the local wall strength, our objective is to determine the maximum stress, and thus the discrepancies in MPS for low percentile ranks are not impacting our analysis. The differences in lower percentile ranks, corresponding to lower stress values, may be attributed to variations in MPS directions between the two models.

Figure 6 demonstrates the validation of our straightforward method for computing wall tension. We validated our method by comparing MPWT computed as the MPS at the shell mid-surface multiplied by the shell thickness, against the MPWT obtained via numerical integration through the shell thickness. In Fig. 6, we compared the MPWT percentile plots obtained from both computation methods and, showing that our straightforward method for MPWT computation yields results nearly identical to those from the numerical integration method.

The contour plot in Fig. 6 shows the MPWT on Patient 1's cerebral aneurysm wall in N/mm, obtained by multiplying the shell thickness by the MPS at the mid-surface. As mentioned previously, in this paper, MPS for shell elements represents the maximum in-plane principal stress output in Abaqus. As shown in Fig. 6, for Patient 1, the maximum wall tension in the cerebral aneurysm is 0.197 N/mm.

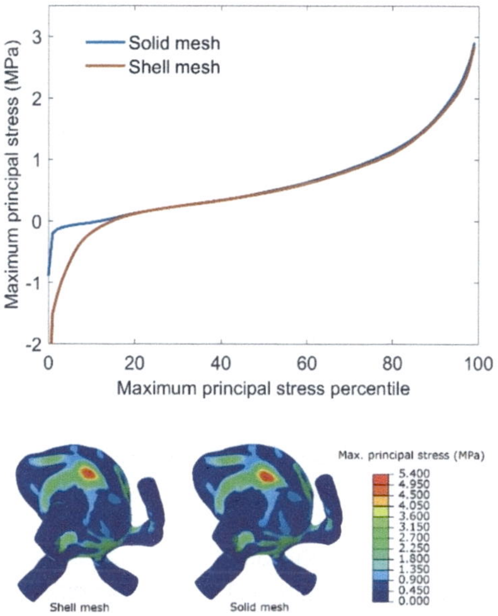

Fig. 5. Verification of the shell finite element mesh by comparing the maximum principal stress (MPS) on the internal surface of the cerebral aneurysm between the shell and solid meshes. MPS contour plots are presented alongside MPS percentile plots to provide a detailed comparison between the solid and shell meshes.

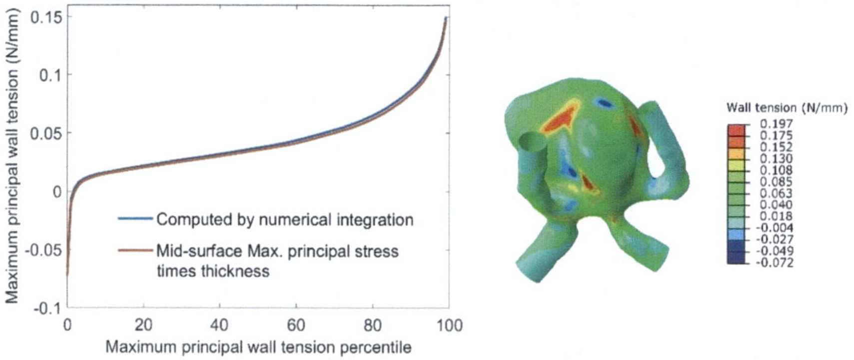

Fig. 6. Verification of the maximum principal wall tension (MPWT), computed as the maximum principal stress (MPS) at the shell mid-surface multiplied by the shell thickness, against MPWT computed via numerical integration through shell thickness. MPWT percentile plots obtained from both computation methods are compared. The contour plot illustrates MPWT on the cerebral aneurysm wall in N/mm obtained by multiplying the shell thickness by MPS at the mid-surface. Here, MPS represents the maximum in-plane principal stress output of shell elements in Abaqus.

To revisit and summarize, we employed our approach to compute cerebral aneurysm wall tension for another patient, referred to as Patient 2, as shown in Fig. 7. Figure 7a shows the surface model of Patient 2's cerebral aneurysm with clipped connecting vessels, extracted from Patient 2's CTA image via image segmentation, centerline extraction, and vessel clipping. Figure 7b shows the FE model using shell elements of size 0.2 mm. Figure 7c shows the contour plot of the MPWT, with a maximum of 0.111 N/mm.

In practice, instead of beginning from scratch, we used a template Abaqus CAE file and updated the template FE model for every new geometry. Specifically, our efficient pipeline involves the following steps, with the surface model of the aneurysm as input: (1) open the existing template Abaqus CAE file, (2) import the new aneurysm geometry, (3) update the existing FE model with the new geometry, (4) update the node set for fixed boundary condition application, (5) update the interior surface for blood pressure application, (6) update the patient-specific blood pressure value, (7) solve the updated FE model using Abaqus/Standard, and (8) create the wall tension field output as thickness multiplied by the maximum in-plane principal stress at the mid-surface. Except for steps (4) and (5) that need user input, the other steps can be automated via Python programming in Abaqus. Since wall thickness has no effect on wall tension, the shell thickness parameter in the template FE model needs no update.

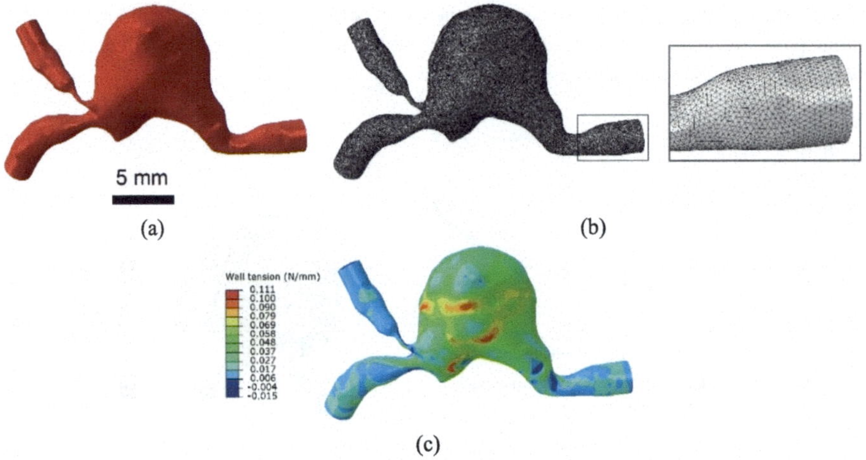

Fig. 7. Patient 2's cerebral aneurysm wall tension computed using the proposed efficient pipeline: (a) Aneurysm geometry, (b) Shell finite element mesh, (c) Contour plot of maximum principal wall tension.

4 Conclusions and Discussions

We presented an efficient streamlined approach for computing patient-specific cerebral aneurysm wall tension using finite element (FE) analysis of an aneurysm biomechanical model. The input to our pipeline is the CTA image of the aneurysm, and the output is the maximum principal wall tension (MPWT) map on the aneurysm wall. Our approach involves three major steps as follows: (1) semi-automatic image segmentation to extract the geometry of cerebral arteries with aneurysm, (2) creating the surface model of the cerebral aneurysm with clipped connecting vessels, and (3) FE analysis of the cerebral aneurysm biomechanical model to extract the MPWT map. Except for image segmentation in step (1), which requires an expert neurologist or radiologist, the other steps can be performed by an analyst with minimal training. Though steps (2) and (3) are not automated due to the variations and complexities in cerebral aneurysm geometry, they are straightforward and quick.

Our methodological innovation is the implementation and verification of an efficient approach for computing MPWT in cerebral aneurysms using shell elements. The shell FE model significantly simplifies the FE model creation and wall tension extraction processes. Our approach eliminates the need for numerical integration through the aneurysm wall thickness for computing MPWT. In our approach, MPWT is easily computed in Abaqus by multiplying the shell thickness by the maximum in-plane principal stress at the mid-surface, which is a direct output from Abaqus. In practice, the FE model creation is simplified by starting from a template Abaqus FE model and updating it for a new aneurysm geometry.

We assumed linear elastic, isotropic material properties for the cerebral aneurysm wall and a constant wall thickness in our analysis. However, these assumptions and simplifications do not compromise the validity of our results. In previous studies on abdominal aortic aneurysms (AAA), we demonstrated that, from the perspective of distinguishing between AAAs that would rupture and those that would remain intact, linear elastic analysis with isotropic material properties yields practically the same results as more complex non-linear models [39]. Furthermore, our simulations with varying lengths of the clipped connecting vessels (not shown here for brevity) indicate that the clipping location has a negligible effect on aneurysm wall tension.

Low image resolution and poor soft tissue contrast impair the accuracy and reliability of artery wall thickness measurements from CT images. As a result, constant wall thickness has been assumed in this and many other studies on cerebral aneurysm biomechanics [11, 12].

In conclusion, our approach allows for the non-invasive computation of patient-specific wall tension without requiring material properties and wall thickness. It can be used to investigate the correlation between aneurysm wall tension and the risk of aneurysmal re-bleeding. Once validated, the pipeline can be utilized in patient-specific disease management, potentially improving patient outcomes through more tailored treatment plans.

References

1. D'Souza, S.: Aneurysmal subarachnoid hemorrhage. J. Neurosurg. Anesthesiol. **27**, 222–240 (2015)
2. Toyoda, K., et al.: Twenty-year change in severity and outcome of ischemic and hemorrhagic strokes. JAMA Neurol. **79**, 61–69 (2022)
3. Rinkel, G.J., Algra, A.: Long-term outcomes of patients with aneurysmal subarachnoid haemorrhage. Lancet Neurol. **10**, 349–356 (2011)

4. Korja, M., Lehto, H., Juvela, S., Kaprio, J.: Incidence of subarachnoid hemorrhage is decreasing together with decreasing smoking rates. Neurology **87**, 1118–1123 (2016)
5. Larsen, C.C., Astrup, J.: Rebleeding after aneurysmal subarachnoid hemorrhage: a literature review. World Neurosurg. **79**, 307–312 (2013)
6. Hoh, B.L., et al.: Guideline for the management of patients with aneurysmal subarachnoid hemorrhage: a guideline from the american heart association/american stroke association. Stroke **54**, e314–e370 (2023)
7. Hadjiathanasiou, A., et al.: Multiple aneurysms in subarachnoid hemorrhage - identification of the ruptured aneurysm, when the bleeding pattern is not self-explanatory - development of a novel prediction score. BMC Neurol. **20**, 70 (2020)
8. Miller, K., et al.: Is there a relationship between stress in walls of abdominal aortic aneurysm and symptoms? J. Surg. Res. **252**, 37–46 (2020)
9. Lauzier, D.C., et al.: Early brain injury after subarachnoid hemorrhage: incidence and mechanisms. Stroke **54**, 1426–1440 (2023)
10. Al-Mufti, F., et al.: Emerging markers of early brain injury and delayed cerebral ischemia in aneurysmal subarachnoid hemorrhage. World Neurosurg. **107**, 148–159 (2017)
11. Ma, B., Lu, J., Harbaugh, R.E., Raghavan, M.L.: Nonlinear anisotropic stress analysis of anatomically realistic cerebral aneurysms. J. Biomech. Eng-T Asme **129**, 88–96 (2007)
12. Raghuram, A., et al.: Comprehensive morphomechanical analysis of brain aneurysms. Acta Neurochir. **165**, 461–470 (2023)
13. Lu, J., Hu, S.H., Raghavan, M.L.: A shell-based inverse approach of stress analysis in intracranial aneurysms. Ann. Biomed. Eng. **41**, 1505–1515 (2013)
14. Seshaiyer, P., Hsu, F.P.K., Shah, A.D., Kyriacou, S.K., Humphrey∗, J.D.: Multiaxial mechanical behavior of human saccular aneurysms. Comput. Methods Biomech. Biomed. Eng. **4**, 281–289 (2001)
15. Lu, J., Zhou, X., Raghavan, M.L.: Inverse method of stress analysis for cerebral aneurysms. Biomech. Model. Mechanobiol. **7**, 477–486 (2008)
16. Yadav, P.N., Singh, G., Chanda, A.: Biomechanical modeling of cerebral aneurysm. Mater. Today Proc. **62**, 3295–3300 (2022)
17. Joldes, G.R., Miller, K., Wittek, A., Forsythe, R.O., Newby, D.E., Doyle, B.J.: BioPARR: a software system for estimating the rupture potential index for abdominal aortic aneurysms. Sci. Rep. **7**, 4641 (2017)
18. https://bioparr.mech.uwa.edu.au/
19. Joldes, G.R., Miller, K., Wittek, A., Doyle, B.: A simple, effective and clinically applicable method to compute abdominal aortic aneurysm wall stress. J. Mech. Behav. Biomed. Mater. **58**, 139–148 (2016)
20. Lu, J., Zhou, X., Raghavan, M.L.: Computational method of inverse elastostatics for anisotropic hyperelastic solids. Int. J. Numer. Meth. Eng. **69**, 1239–1261 (2007)
21. Lu, J., Zhou, X., Raghavan, M.L.: Inverse elastostatic stress analysis in pre-deformed biological structures: demonstration using abdominal aortic aneurysms. J. Biomech. **40**, 693–696 (2007)
22. Biehler, J., Gee, M.W., Wall, W.A.: Towards efficient uncertainty quantification in complex and large-scale biomechanical problems based on a bayesian multi-fidelity scheme. Biomech. Model. Mechanobiol. **14**, 489–513 (2015)
23. Zelaya, J.E., Goenezen, S., Dargon, P.T., Azarbal, A.-F., Rugonyi, S.: Improving the efficiency of abdominal aortic aneurysm wall stress computations. PLoS ONE **9**, e101353 (2014)
24. Miller, K., Lu, J.: On the prospect of patient-specific biomechanics without patient-specific properties of tissues. J. Mech. Behav. Biomed. **27**, 154–166 (2013)
25. Liu, M., Liang, L., Liu, H., Zhang, M., Martin, C., Sun, W.: On the computation of in vivo transmural mean stress of patient-specific aortic wall. Biomech. Model. Mechanobiol. **18**, 387–398 (2019)

26. Joldes, G.R., Noble, C., Polzer, S., Taylor, Z.A., Wittek, A., Miller, K.: A simple method of incorporating the effect of the uniform stress hypothesis in arterial wall stress computations. Acta Bioeng. Biomech. **20**, 59–67 (2018)
27. Fung, Y.C.: What are the residual-stresses doing in our blood-vessels. Ann. Biomed. Eng. **19**, 237–249 (1991)
28. Polzer, S., Bursa, J., Gasser, T.C., Staffa, R., Vlachovsky, R.: A numerical implementation to predict residual strains from the homogeneous stress hypothesis with application to abdominal aortic aneurysms. Ann. Biomed. Eng. **41**, 1516–1527 (2013)
29. Miller, K., et al.: Maximum principal AAA wall stress is proportional to wall thickness. Comput. Biomech. Med. Measure. Models Predict. 43–53 (2018)
30. Miller, K., Wittek, A., Joldes, G.: Re "Biomechanical Assessment Predicts Aneurysm Related Events in Patients with Abdominal Aortic Aneurysm." Eur. J. Vasc. Endovasc. Surg. **61**, 163–164 (2021)
31. Fedorov, A., et al.: 3D slicer as an image computing platform for the quantitative imaging network. Magn. Reson. Imaging **30**, 1323–1341 (2012)
32. https://www.slicer.org/
33. http://www.vmtk.org
34. https://www.3ds.com/products/simulia/abaqus
35. Geuzaine, C., Remacle, J.-F.: Gmsh: a 3-D finite element mesh generator with built-in pre - and post-processing facilities. Int. J. Numer. Meth. Eng. **79**, 1309–1331 (2009)
36. https://gmsh.info/
37. Ramachandran, M., Laakso, A., Harbaugh, R.E., Raghavan, M.L.: On the role of modeling choices in estimation of cerebral aneurysm wall tension. J. Biomech. **45**, 2914–2919 (2012)
38. Alkhatib, F., et al.: Towards full automation of geometry extraction for biomechanical analysis of abdominal aortic aneurysm; neural network-based versus classical methodologies. arXiv (2024)
39. Wittek, A., Alkhatib, F., Vitásek, R., Polzer, S., Miller, K.: On stress in abdominal aortic aneurysm: linear versus non-linear analysis and aneurysm rupture risk. Int. J. Numer. Methods Biomed. Eng. **38**, e3554 (2022)
40. Speelman, L., et al.: Patient-specific AAA wall stress analysis: 99-percentile versus peak stress. Euro. J. Vas. Endovas. Surg. **36**, 668–676 (2008)

Abdominal Aortic Aneurysm Wall Stress: A 7-line Code in MATLAB and a One-Click Software Application

Mostafa Jamshidian[1(✉)], Saeideh Sekhavat[1], Adam Wittek[1], Donatien Le Liepvre[2], Florian Bernard[2], Ludovic Minvielle[2], Antoine Fondanèche[2], and Karol Miller[1]

[1] Intelligent Systems for Medicine Laboratory, The University of Western Australia, Perth, WA, Australia
mostafa.jamshidian@uwa.edu.au
[2] Nurea, Bordeaux, France

Abstract. An abdominal aortic aneurysm (AAA) is a life-threatening condition characterized by the irreversible dilation of the lower aorta, usually detected incidentally during imaging for other health issues. Current clinical practice for managing AAA relies on a one-size-fits-all approach, based on the aneurysm's maximum diameter and growth rate, which can lead to underestimation or overestimation of AAA rupture risk. Patient-specific AAA wall stress, computed using biomechanical models derived from medical images without needing patient-specific material properties, has been widely investigated for developing individualized AAA rupture risk predictors. Therefore, AAA wall stress, determined reliably and quickly, has the potential to enhance patient-specific treatment plans. This paper presents a 7-line code, written in MATLAB using the Partial Differential Equation Toolbox, for AAA wall stress computations via finite element analysis. The code takes AAA wall geometry as input and outputs stress components over the AAA wall domain. Additionally, we present a one-click standalone software application for AAA wall stress computation, developed based on our 7-line code using MATLAB Compiler. After verification, we used our code to compute AAA wall stress in ten patients. Our analysis indicated that the 99th percentile of maximum principal stress across all patients ranged from 0.307 MPa to 0.466 MPa, with an average of 0.380 MPa and a standard deviation of 0.048 MPa. Moreover, for every case, the MATLAB simulation time was less than a minute on a laptop workstation.

Keywords: Abdominal Aortic Aneurism · Biomechanics · Software

1 Introduction

An abdominal aortic aneurysm (AAA) is an asymptomatic, permanent, and irreversible dilation of the lower aorta that is commonly diagnosed incidentally during imaging for other health conditions. If left untreated, AAA can expand to the point of rupture, leading to death most of the time [1, 2].

M. Kobielarz et al. (Eds.): MICCAI 2024, LNBE, pp. 88–103, 2025.
https://doi.org/10.1007/978-3-031-94128-3_9

Currently, clinical practice for AAA management is a one-size-fits-all approach based on the aneurysm's maximum diameter and growth rate. Clinical intervention should be recommended when aneurysm diameters exceed 5.5 cm for men and 5 cm for women, or when the growth rate exceeds 1 cm annually [2]. As demonstrated by ruptured AAAs with diameters smaller than the critical diameter [3] and stable AAAs with diameters larger than the critical diameter [1, 2, 4], the maximum diameter criteria may underestimate or overestimate the rupture risk. Autopsy results indicate that approximately 13% of AAAs with a diameter of 5 cm or less ruptured, while 60% of AAAs larger than 5 cm in diameter did not rupture [5]. It is thus possible to improve lives and reduce healthcare costs by developing a patient-specific AAA rupture risk predictor.

Based on mechanics, an artery ruptures when the local wall stress exceeds the local wall strength. Hence, AAA wall stress, computed based on patient-specific biomechanical models extracted from medical images, has widely been investigated to develop a patient-specific AAA rupture risk predictor [6–15]. Significant progress in wall stress computation has been achieved through the invention of biomechanical models that provide wall tension without necessitating mechanical properties and thickness distribution of the wall [16]. AAA wall stress, determined reliably and quickly through non-invasive approaches, can potentially help with patient-specific treatment.

BioPARR (Biomechanics based Prediction of Aneurysm Rupture Risk) is a free software package for AAA stress analysis based on the finite element (FE) method [17, 18]. Fed with the patient's blood pressure and the AAA and blood label maps extracted from the medical images, BioPARR conducts automatic robust computations based on the recent computational biomechanics approaches that obviate the requirement for patient-specific material properties [16]. As a truly patient-specific approach based solely on clinical data, BioPARR presents a clinically applicable standardized approach for the comparative evaluation of AAA.

While seamless at the time of release, BioPARR can be improved based on recent software advancements. BioPARR relies on multiple libraries and external software including the freely available 3D Slicer image computing platform [19, 20], the scientific visualization free software ParaView, the FE mesh generator free software Gmsh [21, 22], and the commercially-available Abaqus FE software [23]. BioPARR functions optimally with the recommended versions of these external software, but subsequent updates may compromise its robustness. Generally, the recommended older version of freeware may lack adequate support for new hardware and operating systems. Hence, BioPARR requires updates and maintenance, aligned with updates to the external software. Another limitation involves using expensive commercial software like Abaqus, which could potentially be substituted with freely available FE packages.

In this paper, we introduce a 7-line code and a one-click software application for AAA stress analysis written in MATLAB programming language. The remainder of the paper is organized as follows: In Sect. 2, we present the methods for AAA stress analysis in MATLAB and its verification against BioPARR stress analysis using Abaqus FE software [23]. We also describe the one-click software application developed using the 7-line code as the backend. In Sect. 3, we present the results for the mesh convergence analysis in MATLAB and verification of the MATLAB FE analysis. We also present the

AAA stress analysis in ten patients. Finally, Sect. 4 is devoted to the conclusions and discussions.

2 Methods

2.1 Image Data and AAA Surface Model

We employed anonymized contrast-enhanced 3D CTA images of ten patients diagnosed with AAA. These patients were recruited at Fiona Stanley Hospital in Western Australia and provided informed consent before participating in the study. The research was conducted in accordance with the Declaration of Helsinki, and the protocol received approval from the Human Research Ethics and Governance at South Metropolitan Health Service (HREC-SMHS) under approval code RGS3501, as well as from the Human Research Ethics Office at The University of Western Australia under approval code RA/4/20/5913.

As an example of the image data, Fig. 1a displays the cropped 3D CTA image of Patient 1's AAA. This image has dimensions of (112, 109, 74) voxels and voxel spacing of (0.627, 0.627, 1.000) mm along the (R, A, S) axes, as shown in Fig. 1a. In the patient coordinate system, the basis aligns with the anatomical axes of anterior-posterior, inferior-superior, and left-right. Specifically, the R, A, and S axes correspond to the left-right, posterior-anterior, and inferior-superior directions, respectively.

To create the patient's AAA surface model used in FE analysis, we used PRAEVAorta by NUREA (https://www.nurea-soft.com/) for AI-based automatic segmentation of AAA image in Fig. 1a, followed by automated post-processing with in-house MATLAB code [24], to automatically extract the AAA and blood label maps, as shown in Fig. 1b and Fig. 1c, respectively. We then used these label maps and the assumed wall thickness of 1.5 mm [25] as input for BioPARR to automatically extract the wall geometry and create the AAA surface model, as shown in Fig. 1d. It should be noted that the AAA surface model shown in Fig. 1d does not contain an intraluminal thrombus (ILT). We saved the surface model as an STL file, which was used as the input geometry for the MATLAB FE analysis.

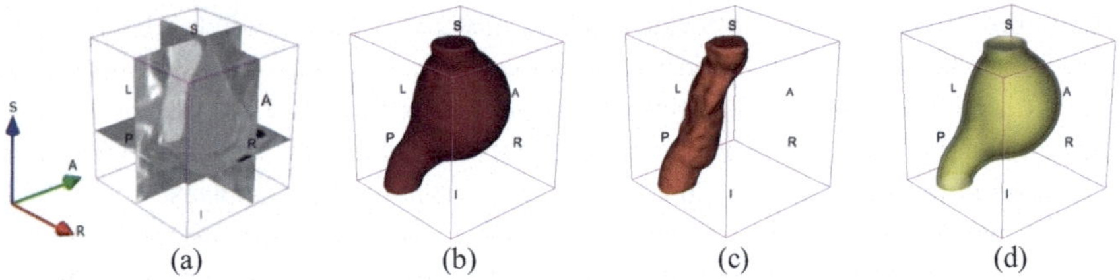

Fig. 1. Patient 1's abdominal aortic aneurysm (AAA): (a) Cropped 3D CTA image, (b) AAA label map, (c) Blood label map, and (d) Surface model of AAA wall.

2.2 Stress Analysis in MATLAB

The medical image data captures the geometry of a pressurized artery deformed by blood pressure. As a result, the artery geometries reconstructed from these images are the deformed configurations of the vessel structure. In biomechanics problems involving deformed geometries, material properties have minimal impact on wall stress [26, 27]. Consequently, we calculated the patient-specific AAA wall stress using a direct linear analysis without incorporating patient-specific material property data [16].

Starting from the AAA surface model shown in Fig. 1d, we conducted the stress analysis using MATLAB Partial Differential Equation (PDE) Toolbox. We performed AAA wall stress analysis in MATLAB via a 7-line code as follows:

```
structuralModel = createpde("structural","static-solid");
importGeometry(structuralModel,'WallSurface.stl');
structuralProper-
ties(structuralModel,"Cell",1,"YoungsModulus",100000,"PoissonsRatio",0.49);
structuralBC(structuralModel,"Face",[3,4],"Constraint","fixed");
structuralBoundaryLoad(structuralModel,"Face",1,"Pressure",13e-03);
generateMesh(structuralModel,"Hmax",1.0,'GeometricOrder','quadratic');
structuralresults = solve(structuralModel);
```

Line 1 creates a static structural model for solving a small-strain linear elasticity three-dimensional problem.

Line 2 creates a geometry object from the STL geometry file "WallSurface.stl" of the AAA wall surface model shown in Fig. 1d. The imported geometry object in MATLAB is shown in Fig. 2a. As shown in Fig. 2a, MATLAB automatically detects the cells and the faces of the imported geometry. For the AAA geometry, MATLAB detects one cell as C1 for the whole AAA wall geometry and four faces including the interior surface as F1, the exterior surface as F2, the bottom flat surface as F3, and the top flat surface as F4, as shown in Fig. 2a. We used these face labels to apply loads and boundary conditions.

Line 3 assigns isotropic linear elastic material properties for the structural model using Young modulus of 100 GPa and Poisson's ratio of 0.49. This high stiffness minimizes further deformation of the already deformed geometry of the AAA, with results remaining independent of Young modulus in stress recovery under a force-controlled scenario.

Line 4 applies the boundary conditions to the structural model by fixing the bottom (face F3) and the top (face F4) flat surfaces.

Line 5 applies a uniform pressure on the interior surface (face F1) of the aneurysm geometry using the patient-specific blood pressure of 13 kPa.

Line 6 meshes the geometry with quadratic tetrahedral elements using the target maximum element edge length of 1.0 mm, as shown in Fig. 2b.

Line 7 solves the FE model and returns the results including displacement, strain, and stress components at all nodes in the FE mesh. These primary results can be further processed to determine any desired quantities, such as maximum principal stress (MPS).

As an example of the MATLAB FE analysis results, Fig. 2c shows the MPS contour plot on Patient 1's AAA wall.

For MATLAB FE analysis, we performed FE mesh convergence analysis using three different element sizes of 1.5, 1.0, and 0.5 mm, as shown in Fig. 3. The coarse FE mesh in Fig. 3a with FE size of 1.5 mm consists of 59962 nodes and 30364 elements. The FE mesh with medium mesh refinement using an FE size of 1.0 mm in Fig. 3b consists of 182506 nodes and 106675 elements. The fine FE mesh in Fig. 3c with an FE size of 0.5 mm consists of 1059351 nodes and 683966 elements.

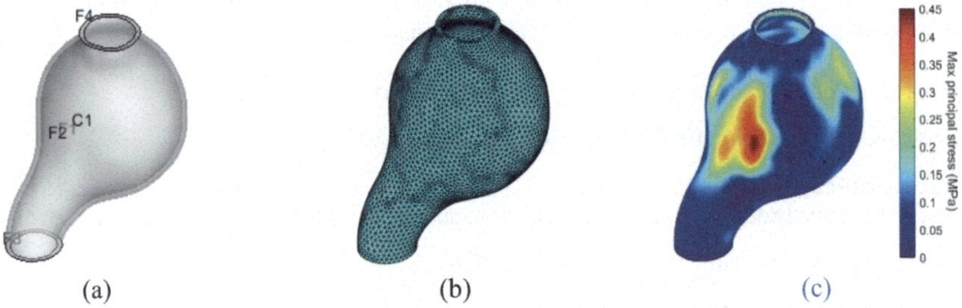

Fig. 2. Finite element (FE) stress analysis of abdominal aortic aneurysm (AAA) wall in MATLAB: (a) Surface model imported in MATLAB, (b) MATLAB FE Mesh, (c) MATLAB FE analysis results for the maximum principal stress (MPS) on AAA wall.

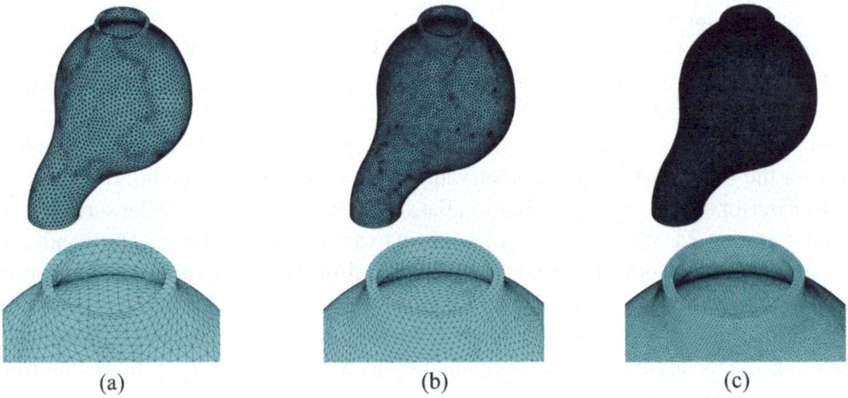

Fig. 3. Finite element (FE) mesh convergence Analysis in MATLAB using three different element sizes including (a) FE size 1.5 mm, (b) FE size 1.0 mm, (c) FE size 0.5 mm. Mesh refinement increases from left to right, with coarser meshes on the left and finer meshes on the right.

2.3 Verification Against Abaqus

Alkhatib et al. [28] investigated the effects of FE formulation, FE shape functions, spatial integration scheme, and number of elements through the wall thickness, on the efficiency, robustness, and accuracy of AAA wall stress computations. They showed that

Table 1. AAA wall stress analysis results for ten patients using MATLAB. For each patient, the AAA wall geometry, and the contour plots of maximum principal stress (MPS) are displayed, with the 99th percentile of MPS used as the maximum contour limit.

Patient number	AAA wall geometry	Maximum principal stress (MPa)		
1	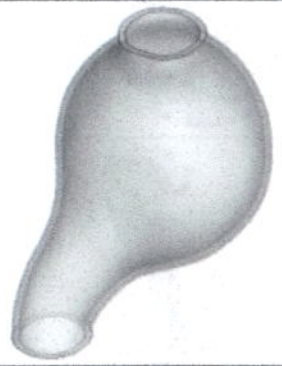	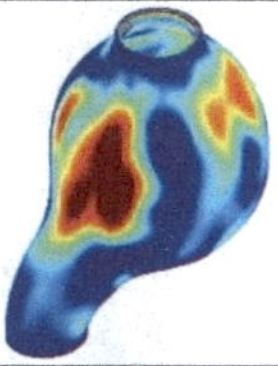	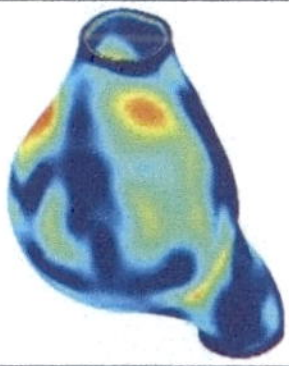	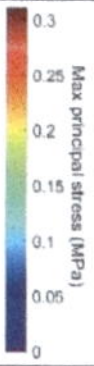
2		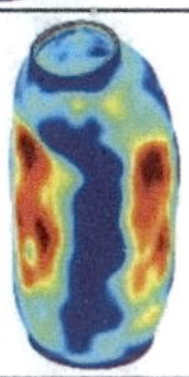	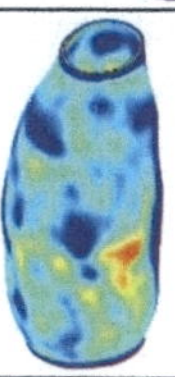	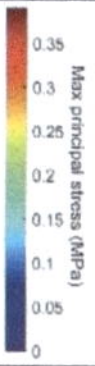
3		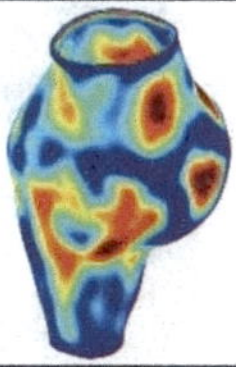	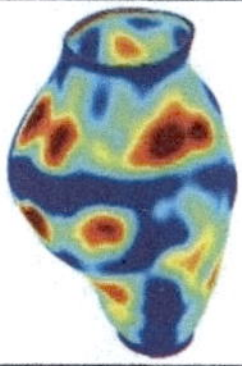	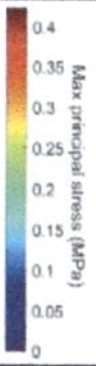
4	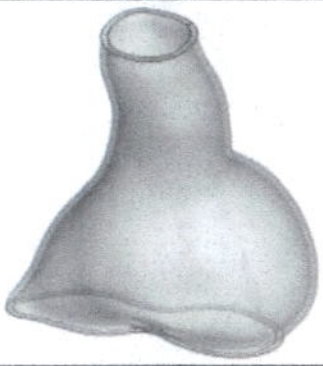	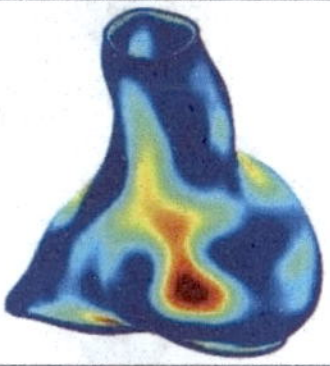	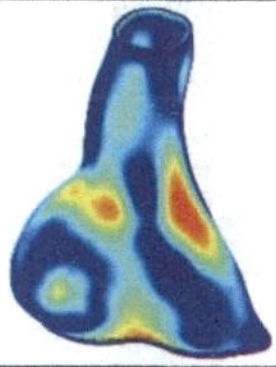	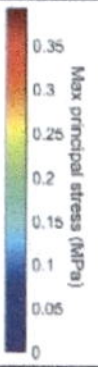
5		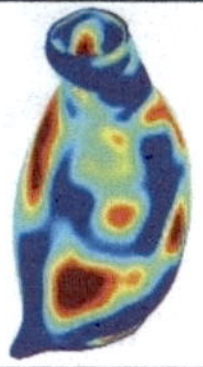	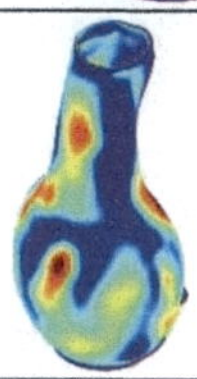	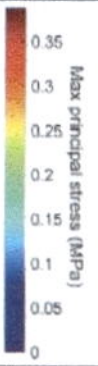

(continued)

Table 1. (*continued*)

6		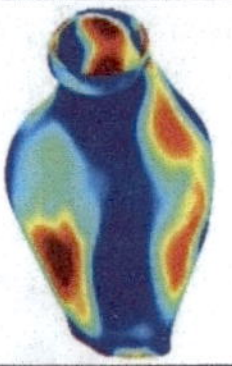	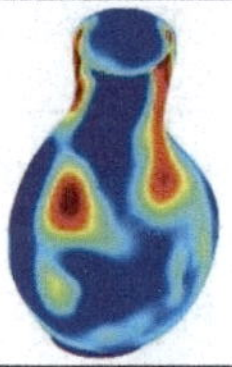	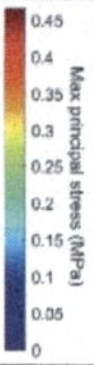
7	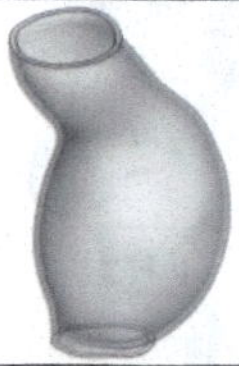	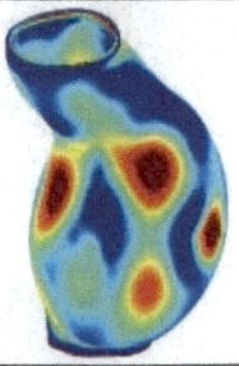	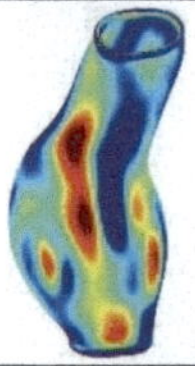	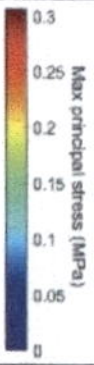
8	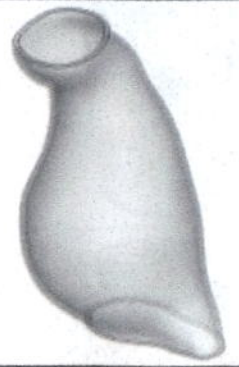	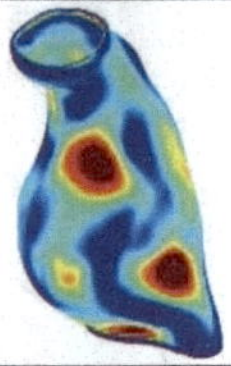	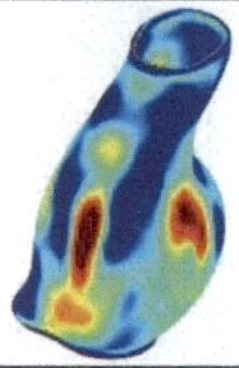	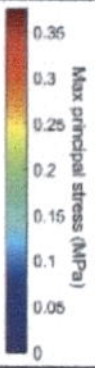
9	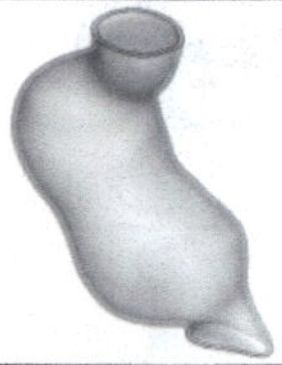	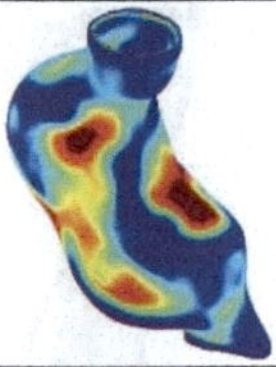	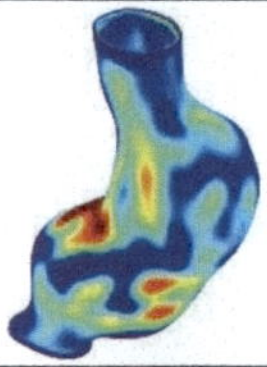	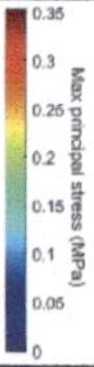
10		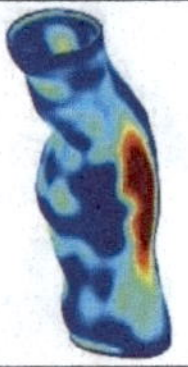	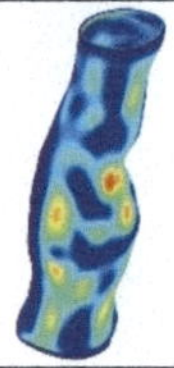	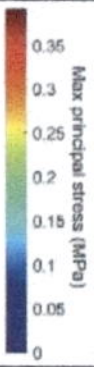

Table 2 indicates that the MPS 99th percentile among all patients ranged from 0.307 MPa to 0.466 MPa, with an average of 0.380 MPa and a standard deviation of 0.048 MPa, which is 13% of the average. This relatively low standard deviation-to-average ratio suggests consistency in AAA wall stress across all patients.

Table 2 also shows that the average MATLAB simulation time for AAA stress analyses is 38 s, with a maximum of 45 s and a minimum of 29 s. This demonstrates that the entire computations, from STL geometry to AAA wall stress, can be completed in under 1 min.

Table 2. AAA wall stress analysis results for ten patients using MATLAB. For each patient, the 99th percentile of maximum principal stress (MPS) on the AAA wall and the finite element (FE) simulation time in MATLAB are reported.

Patient number	MPS 99th percentile (MPa)	Simulation time (s)
1	0.311	36
2	0.391	41
3	0.424	43
4	0.393	29
5	0.388	45
6	0.466	32
7	0.307	35
8	0.376	42
9	0.354	40
10	0.392	37
Minimum	0.307	29
Maximum	0.466	45
Average	0.380	38
Standard deviation	0.048	5

4 Conclusions and Discussions

We introduced a 7-line MATLAB code utilizing the Partial Differential Equation (PDE) Toolbox for computing AAA wall stress through finite element (FE) analysis. Additionally, using MATLAB Compiler, we developed a standalone one-click software application for AAA wall stress computation, with our 7-line code running in the backend. The input to our 7-line code is the AAA wall surface model derived from medical images, and its output is the stress components over the AAA wall. Our code utilizes a stress computation method, previously implemented in the free software package BioPARR (Biomechanics based Prediction of Aneurysm Rupture Risk), that eliminates the need for patient-specific material properties.

We performed a mesh convergence analysis for our MATLAB FE analysis and demonstrated that a tetrahedral mesh with quadratic elements of size 1.0 mm provides sufficiently accurate results for the maximum principal stress (MPS). We verified our MATLAB FE analysis against previously validated Abaqus FE analysis.

We applied our 7-line code to compute AAA wall stress in ten patients. The FE analysis results indicated that the 99th percentile of MPS across all patients ranged from 0.307 MPa to 0.466 MPa, with an average of 0.380 MPa and a standard deviation of 0.048 MPa. Furthermore, the entire MATLAB simulation time for each case was under a minute using a laptop workstation.

We assumed linear elastic, isotropic material properties for the AAA wall, constant wall thickness, and did not include residual stress or intraluminal thrombus (ILT) in our analysis. However, these assumptions and simplifications do not compromise the validity of our results. We have previously shown that, from the perspective of distinguishing between AAAs that would rupture and those that would remain intact, linear elastic analysis using isotropic material properties yields practically the same results as more complicated non-linear models [30].

Low image resolution and poor soft tissue contrast impair the accuracy and reliability of aortic wall thickness measurements from CT images. Consequently, in this and many other studies of AAA biomechanics, constant wall thickness has been used [3, 10, 33]. BioPARR [17, 18] utilizes a very efficient method for computing residual stress based on the uniform stress hypothesis by Fung [34]. This method could have been incorporated into the approach and computer code presented in this study. However, including residual stress makes the distribution of stress within the AAA wall more uniform, which would be inconsistent with the purpose of verifying AAA stress determined using our fully automated approach against the results obtained using procedures available in established general-purpose finite element codes (we used Abaqus). Therefore, in this study, we did not account for the effects of residual stress.

The analysis of the effects of intraluminal thrombus is beyond the scope of this study. However, the automated approach for creating patient-specific AAA surface models from CT images, described in Sect. 2.1 (Image Data and AAA Surface Model), has been previously successfully applied to AAAs with ILT [24].

One major limitation of our method is the need for wall strength to determine stress-based rupture risk indicators [35, 36]. Since it is clinically impractical to extract in-vivo wall strength, population-based statistics are used to estimate wall strength, which do not accurately reflect individual patient conditions [37]. Recent research suggests that the dimensionless ratio of wall stress to wall strength at baseline is a predictor of AAA rupture or repair, improving risk stratification for AAA events compared to solely using the maximum diameter [35, 36].

However, the fact that the material properties of the wall vary among patients and locally within an individual's aortic wall significantly affects rupture risk estimation [38, 39]. Obviously, direct non-invasive measurement of in-vivo local patient-specific wall strength (or degradation) in a living individual is impossible, but it may be indirectly estimated by supplementing the wall stress map with the wall strain map [40]. Towards a more predictive rupture risk estimation, there have been attempts at non-invasive, in-vivo measurement of AAA wall strain using sequential images at different phases of the cardiac cycle [40–43].

Despite these limitations, our 7-line MATLAB code and standalone software application offer the clear advantage of fast and reliable AAA wall stress computation.

Acknowledgements. This work was supported by the Australian National Health and Medical Research Council NHMRC Ideas grant no. APP2001689. The authors acknowledge the contributions of Christopher Wood and Jane Polce, radiology technicians at the Medical Imaging Department, Fiona Stanley Hospital (Murdoch, Western Australia), to patient image acquisition, and Dr. Farah Alkhatib's (The University of Western Australia) assistance in patient recruitment and image acquisition.

References

1. National Institute for Health and Care Excellence. www.nice.org.uk/guidance/ng156
2. Wanhainen, A., et al.: Editor's Choice – European Society for Vascular Surgery (ESVS) 2019 Clinical Practice Guidelines on the Management of Abdominal Aortoiliac Artery Aneurysms. Euro. J. Vascular Endo. Surg. 57, 8–93 (2019)
3. Vorp, D.A.: Biomechanics of abdominal aortic aneurysm. J. Biomech. **40**, 1887–1902 (2007)
4. Darling, R., Messina, C., Brewster, D., Ottinger, L.: Autopsy study of unoperated abdominal aortic aneurysms. The case for early resection. Circulation **56**, II161–164 (1977)
5. Kontopodis, N., Pantidis, D., Dedes, A., Daskalakis, N., Ioannou, C.V.: The - Not So - Solid 5.5 cm threshold for abdominal aortic aneurysm repair: facts, misinterpretations, and future directions. Front Surg. **3**, 1 (2016)
6. Fillinger, M.F., Raghavan, M.L., Marra, S.P., Cronenwett, J.L., Kennedy, F.E.: In vivo analysis of mechanical wall stress and abdominal aortic aneurysm rupture risk. J. Vasc. Surg. **36**, 589–597 (2002)
7. Fillinger, M.F., Marra, S.P., Raghavan, M.L., Kennedy, F.E.: Prediction of rupture risk in abdominal aortic aneurysm during observation: wall stress versus diameter. J. Vasc. Surg. **37**, 724–732 (2003)
8. Vande Geest, J., Di Martino, E., Bohra, A., Makaroun, M.S., Vorp, D.A.: A biomechanics-based rupture potential index for abdominal aortic aneurysm risk assessment: demonstrative application. Ann. N. Y. Acad. Sci. **1085**, 11–21 (2006)
9. Gasser, T.C., Auer, M., Labruto, F., Swedenborg, J., Roy, J.: Biomechanical rupture risk assessment of abdominal aortic aneurysms: model complexity versus predictability of finite element simulations. Eur. J. Vasc. Endovasc. Surg. **40**, 176–185 (2010)
10. Miller, K., et al.: Is there a relationship between stress in walls of abdominal aortic aneurysm and symptoms? J. Surg. Res. **252**, 37–46 (2020)
11. Indrakusuma, R., et al.: Biomechanical imaging markers as predictors of abdominal aortic aneurysm growth or rupture: a systematic review. Eur. J. Vasc. Endovasc. Surg. **52**, 475–486 (2016)
12. Wang, X.C., et al.: A review on the biomechanical behaviour of the aorta. J. Mech. Behav. Biomed. **144** (2023)
13. Gasser, T.C., Miller, C., Polzer, S., Roy, J.: A quarter of a century biomechanical rupture risk assessment of abdominal aortic aneurysms. Achievements, clinical relevance, and ongoing developments. Int. J. Numer. Meth. Bio (2022)
14. Polzer, S., et al.: Biomechanical indices are more sensitive than diameter in predicting rupture of asymptomatic abdominal aortic aneurysms. J. Vasc. Surg. **71**, 617-+ (2020)
15. Farotto, D., Segers, P., Meuris, B., Sloten, J.V., Famaey, N.: The role of biomechanics in aortic aneurysm management: requirements, open problems and future prospects. J. Mech. Behav. Biomed. **77**, 295–307 (2018)
16. Joldes, G.R., Miller, K., Wittek, A., Doyle, B.: A simple, effective and clinically applicable method to compute abdominal aortic aneurysm wall stress. J. Mech. Behav. Biomed. **58**, 139–148 (2016)
17. Joldes, G.R., Miller, K., Wittek, A., Forsythe, R.O., Newby, D.E., Doyle, B.J.: BioPARR: a software system for estimating the rupture potential index for abdominal aortic aneurysms. Sci. Rep. **7**, 4641 (2017)
18. https://bioparr.mech.uwa.edu.au/
19. Fedorov, A., et al.: 3D slicer as an image computing platform for the quantitative imaging network. Magn. Reson. Imaging **30**, 1323–1341 (2012)
20. https://www.slicer.org/

21. Geuzaine, C., Remacle, J.-F.: Gmsh: a 3-D finite element mesh generator with built-in pre - and post-processing facilities. Int. J. Numer. Meth. Eng. **79**, 1309–1331 (2009)
22. https://gmsh.info/
23. https://www.3ds.com/products/simulia/abaqus
24. Alkhatib, F., et al.: Towards Full Automation of Geometry Extraction for Biomechanical Analysis of Abdominal Aortic Aneurysm; Neural Network-Based versus Classical Methodologies. arXiv preprint arXiv:2403.07238v2 (2024)
25. Raut, S.S., Jana, A., De Oliveira, V., Muluk, S.C., Finol, E.A.: The importance of patient-specific regionally varying wall thickness in abdominal aortic aneurysm biomechanics. J. Biomech. Eng. **135**, 081010 (2013)
26. Miller, K., Lu, J.: On the prospect of patient-specific biomechanics without patient-specific properties of tissues. J. Mech. Behav. Biomed. **27**, 154–166 (2013)
27. Lu, J., Zhou, X.L., Raghavan, M.L.: Inverse elastostatic stress analysis in pre-deformed biological structures: demonstration using abdominal aortic aneurysms. J. Biomech. **40**, 693–696 (2007)
28. Alkhatib, F., Wittek, A., Zwick, B.F., Bourantas, G.C., Miller, K.: Computation for biomechanical analysis of aortic aneurysms: the importance of computational grid. Comput. Methods Biomech. Biomed. Eng. 1–17 (2023)
29. Alkhatib, F., Bourantas, G.C., Wittek, A., Miller, K.: Generation of patient-specific structured hexahedral mesh of aortic aneurysm wall. In: Nash, M.P., Wittek, A., Nielsen, P.M.F., Kobielarz, M., Babu, A.R., Miller, K. (eds.) Computational Biomechanics for Medicine. MICCAI 2021. Springer, Cham (2023). https://doi.org/10.1007/978-3-031-34906-5_1
30. Wittek, A., Alkhatib, F., Vitásek, R., Polzer, S., Miller, K.: On stress in abdominal aortic aneurysm: linear versus non-linear analysis and aneurysm rupture risk. Int. J. Numer. Methods Biomed. Eng. **38**, e3554 (2022)
31. Speelman, L., et al.: Patient-specific AAA wall stress analysis: 99-percentile versus peak stress. Euro. J. Vas. Endovas. Surg. **36**, 668–676 (2008)
32. https://altair.com/hypermesh
33. Inzoli, F., Boschetti, F., Zappa, M., Longo, T., Fumero, R.: Biomechanical factors in abdominal aortic aneurysm rupture. Eur. J. Vasc. Surg. **7**, 667–674 (1993)
34. Fung, Y.C.: What are the residual stresses doing in our blood vessels? Ann. Biomed. Eng. **19**, 237–249 (1991)
35. Singh, T.P., et al.: Association between aortic peak wall stress and rupture index with abdominal aortic aneurysm-related events. Eur. Radiol. **33**, 5698–5706 (2023)
36. Doyle, B.J., et al.: Biomechanical assessment predicts aneurysm related events in patients with abdominal aortic aneurysm. Eur. J. Vasc. Endovasc. Surg. **60**, 365–373 (2020)
37. Miller, K., Wittek, A., Joldes, G.: Re "biomechanical assessment predicts aneurysm related events in patients with abdominal aortic aneurysm." Eur. J. Vasc. Endovasc. Surg. **61**, 163–164 (2021)
38. Li, Z.Y., U-Ying-Im, J., Tang, T.Y., Soh, E., See, T.C., Gillard, J.H.: Impact of calcification and intraluminal thrombus on the computed wall stresses of abdominal aortic aneurysm. J. Vasc. Surg. **47**, 928–935 (2008)
39. Speelman, L., et al.: Effects of wall calcifications in patient-specific wall stress analyses of abdominal aortic aneurysms. J. Biomech. Eng-T Asme **129**, 105–109 (2007)
40. Nagy, R., Csobay-Novák, C., Lovas, A., Sótonyi, P., Bojtár, I.: Non-invasive time-dependent strain measurement method in human abdominal aortic aneurysms: towards a novel approach to rupture risk estimation. J. Biomech. **48**, 1876–1886 (2015)
41. Raut, S.S., Jana, A., Muluk, S.C., Doyle, M., Biederman, R.W., Finol, E.A.: Estimation of patient-specific 3d in vivo abdominal aortic aneurysm strain. In: Proceedings of the Asme Summer Bioengineering Conference 2013, Pt B (2014)

42. Satriano, A., Rivolo, S., Martufi, G., Finol, E.A., Di Martino, E.S.: In vivo strain assessment of the abdominal aortic aneurysm. J. Biomech. **48**, 354–360 (2015)
43. Wittek, A., Blase, C., Derwich, W., Schmitz-Rixen, T., Fritzen, C.P.: Characterization of the mechanical behaviour and pathophysiological state of abdominal aortic aneurysms based on 4D ultrasound strain imaging. Proc. SPIE **10333** (2017)

Markerless Tracking of the Skin Surface

Robin Laven[1(✉)], Gonzalo Maso Talou[1], Alexander Dixon[1], Martyn P. Nash[1,2], Thiranja P. Babarenda Gamage[1], and Poul M. F. Nielsen[1,2]

[1] Auckland Bioengineering Institute, University of Auckland, Auckland, New Zealand
r.laven@auckland.ac.nz

[2] Department of Engineering Science and Biomedical Engineering, University of Auckland, Auckland, New Zealand

Abstract. Biological tissues exhibit complicated motions, like stretching and sliding in-plane, that can have limited impact on the shape of the tissue. These may make identifying constitutive behaviours more difficult. We provide a method for embedding sparse texture-based features into mathematical models, and track an induced deformation in porcine skin using a five camera stereoscope. However we demonstrate that, although these features improve the recovery of the surface motion, sparse features are weak constraints on the surrounding tissues and allow error to accumulate. As such we question what can be inferred from sparse tracking methods, and suggest that augmentation with dense tracking may be necessary for high precision tracking of biological tissues.

1 Introduction

Tracking the surface of biological tissues is crucial for measuring the boundary deformations that allow the identification of material properties. This tracking typically involves understanding the 3D motion of tissues from 2D images acquired from multiple cameras. Such knowledge is essential for applications like prosthesis design, stress and strain analysis, and assessment of active material properties - like muscular contraction of cardiac trabeculae. Even without further analysis, surface tracking also provides up-to-date information about the current state of an object, which can be used to plan and inform surgical interventions.

However, soft biological tissues are typically difficult to track because they lack common textures and are usually smooth surfaces. When tracking the face, the eyes, lips, and brows, algorithms can follow prominent textures common to all faces, while the nose, ears and chin provide rigid shape-based landmarks. Soft biological tissues typically have limited landmarks, like the nipple of the breast, and can have very limited textures, such as creasing and blemishes. To track biological tissues, the literature describes the use of both sparse texture-based features [1], and dense representations of the object texture [2]. Dense representations typically struggle from the extremely non-convex behaviour common in image registration, however sparse methods have orders of magnitude fewll/Aer control points, and must rely on regularisation to recover smooth motion [1].

M. Kobielarz et al. (Eds.): MICCAI 2024, LNBE, pp. 104–113, 2025.
https://doi.org/10.1007/978-3-031-94128-3_10

Typically, the accuracy of these methods is also explored with sparse measurements of the ground truth, e.g. from an EM position sensor[1] or sparsely annotated points [3].

In this article, we investigate the accuracy achievable with sparse feature based methods as skin undergoes a large non-rigid deformation, using both a sparse marker-based ground truth and image registration techniques [4].

In the literature, the following methods have been considered when addressing the challenges of tracking the skin surface: marker and markerless tracking methods; how the cameras used to image the scene are modelled; and the reconstruction algorithms used to recreate the scene.

Marker Based Tracking. Fiducial markers provide additional landmarks over the surface of interest, adding high quality correspondences[5]. Between times, each marker provides an absolute correspondence, minimising the chance that the tracked surface will drift over time. Detecting these fiducial markers is fast, robust, and accurate, even over many times and between imaging modalities [3]. While surface marker based methods are simple to implement and accurate, they are not applicable in every situation.

Markerless Tracking. In many cases markerless methods are either more convenient or necessary (e.g. in keyhole liver surgery [1]), where there is no preoperative access to the tracking target. Markerless tracking generally addresses either finding the motion between adjacent times, or identifying the same point-of-interest between arbitrary frames. Optical flow algorithms have been extensively applied to biological tissues [6], and range from classical to neural network-based. However phase-based approaches have been shown to work well on biological tissues [4], although they struggle with large motions, tissue deformation, and affine transformations common when imaging in biological tissues.

Feature based methods create direct matches between points in image space, which typically make them robust to large motion and changing positions. These include classical features such as SIFT (scale invariant feature transforms)[7] and learnt features like DISK (DIScrete Keypoints) [8]. Extracting 3D information requires matching features between images, e.g. LightGlue, which uses a transformer based architecture to perform matching [9].

Camera Modelling and Bundle Adjustment. Cameras are typically modelled as pinholes, representing their projective behaviour as a linear projection and then correcting for non-linear distortions [10]. The provably optimal process of recovering the parameters of these cameras is known as bundle adjustment[11]. While typically used to calibrate cameras, bundle adjustment can also be used to make measurements in 3D space from 2D observations.

Reconstruction. Stereo reconstruction is the process of matching information between cameras, and using this to recover a depth estimate for every pixel of the input images. With two cameras, it is assumed that the cameras see almost the same image. This allows the use of rectified images, and matching along epipolar lines [12] with correlation based costs, like ZNCC (zero-mean normalised cross correlation) [1]. Many camera reconstructions cannot assume similar views between all cameras, and a multi-view stereo algorithm must determine which views observe shared features. This process is accelerated using massively parallel algorithms, locally planar models, and adaptive matching regions [13].

2 Methods

We evaluated the ability of SIFT features matched with LightGlue to track the deformation of porcine skin excised from the anterior chest. The tissue was obtained in accordance with the University of Auckland Code of Ethical Conduct for the Use of Animals for Teaching and Research.

2.1 Instrumentation Design

Imaged closely, skin has a prominent texture, which is largely self similar. To provide sufficient detail, this requires a resolution of approximately 50 μm per pixel. We used a close range (300 mm operating distance) camera system composed of five inwards facing 5 MP BFS-U3-50S4M-C monochromatic cameras with 12.5 mm focal length lenses. This close range increases motion blur and shortens depth of field, which was mitigated with short exposures (1 ms), a high f-stop, and bright lighting (24000 Lumens in close proximity). All 5 cameras were jointly calibrated with a bundle adjustment [11]. Each time of the deformation was reconstructed using APD-MVS [13]. This resulted in a series of point clouds C_t (Fig. 1).

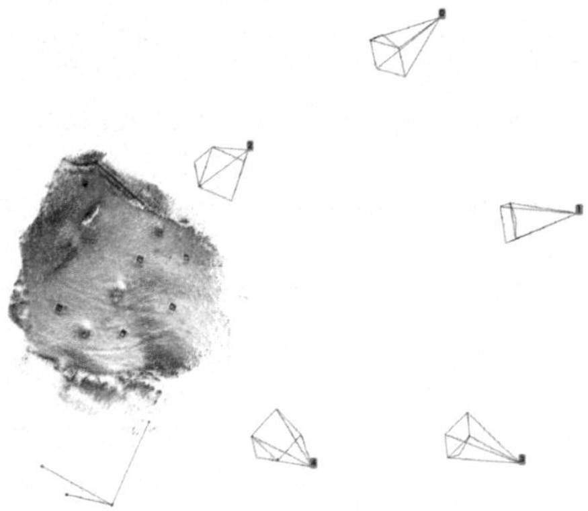

Fig. 1. A representation of the mathematical model of the set of five cameras used to reconstruct the pig skin surface. A frame of the reconstructed pig skin is displayed with reference to the cameras.

2.2 Experimental Design

We assess the results from porcine skin, recorded during a deformation induced by pushing the back surface of the skin forward with a spatula. While the deformation was recorded at 25 frames per second, it was temporally decimated to a 2.5 frames per second to test stability over long-range matches.

2.3 Shape Based Surface Tracking

We used a continuous bi-cubic Hermite mesh to model the central region of porcine skin. This was initialised to the first reconstructed frame by picking four points in the initial point cloud. The parallelogram defined by these four points was subdivided into 9 elements, forming the initial mesh with 192°C of freedom.

This mesh was fitted by minimising the squared distance between a discretised surface representation and the closest point in the reconstructed point cloud C_t, referred to as the surface-to-point distance.

The fit was regularised with a Sobolov-based term penalising non-zero surface derivatives [14]. Given the embedding function ψ, which maps material coordinates ξ to the spatial coordinates $\mathbf{x}$ given the n (192) mesh parameters $\mathbf{P}_s$:

$$\psi : \mathbb{R}^2 \times \mathbb{R}^n \rightarrow \mathbb{R}^3, \psi(\xi, \mathbf{P}_s) = \mathbf{x} \tag{1}$$

The Sobolov smoothing term is given as:

$$\begin{aligned} S(\xi, \mathbf{P}_{st}) =& \alpha_0 || \frac{\partial \psi}{\partial \chi_1}(\xi, \mathbf{P}_{st})||_2 + \alpha_1 || \frac{\partial \psi}{\partial \chi_2}(\xi, \mathbf{P}_{st})||_2 + \alpha_2 || \frac{\partial \psi^2}{\partial \chi_1^2}(\xi, \mathbf{P}_{st})||_2 \\ &+ \alpha_3 || \frac{\partial \psi^2}{\partial \chi_2^2}(\xi, \mathbf{P}_{st})||_2 + \alpha_4 || \frac{\partial \psi^2}{\partial \chi_1 \partial \chi_2}(\xi, \mathbf{P}_{st})||_2 \end{aligned} \tag{2}$$

where the terms χ_n are the directions of the material coordinate space, with weightings $\alpha_n = 1, 1, 0.1, 0.1, 0.1$. Given d as the minimum distance between a single point $\mathbf{x}$ and a set of points, the full loss can be written as:

$$\mathbb{P}_s = \underset{\mathbf{P}_s}{\text{argmin}} \left(\alpha \sum_t^T \sum_e^E S(e, \mathbf{P}_{s,t})^2 + (1 - \alpha) \sum_t^T \sum_e^E d(\psi(e, \mathbf{P}_{s,t}), C_t)^2 \right) \tag{3}$$

where E is the set of points forming the discretised surface representation, and T is the number of imaging times. This was fitted with the weighting $\alpha = 0.1$.

2.4 Feature Detection

As the cameras had large differences in their up-axis, features were detected with the rotationally robust SIFT feature detector, matched with the LightGlue neural-network based matcher [9]. An illustrative representation of the matches is shown in Fig. 2. These features were first matched between all camera images at each time, creating a subset of successfully matched features. The successfully matched features were then matched between all times. These features were reconstructed by bundle adjustment, then filtered based on their epipolar consistency, distance from the reconstruction, and a RANSAC-based rigid motion estimate.

Each feature f was associated with a 3D point in world coordinates, $\mathscr{W}$, $\mathbf{x}_{f,t} \in \mathscr{W}$ in each time t.

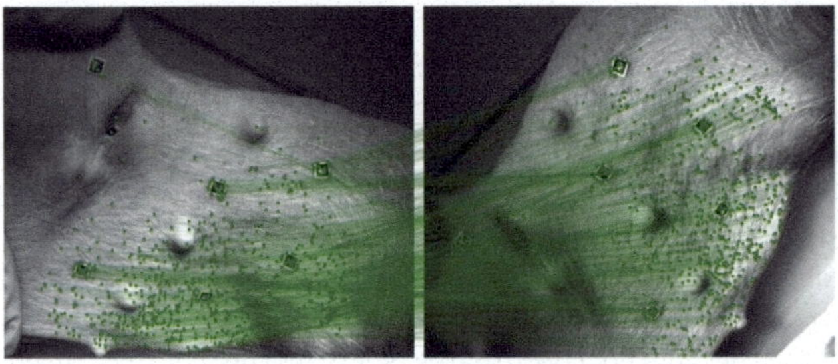

Fig. 2. Recovered matches between two cameras at a single time-step. The matches were obtained using SIFT features then paired with LightGlue. The coverage of the target is generally good, but the combination of SIFT+LightGlue struggled to find consistent matches in some regions.

2.5 Feature Embedding

To embed a feature, we find an associated, constant, material point ξ_f. We assign the initial coordinate ξ_f by finding the closest mesh material coordinate ξ to $\mathbf{x}_{f,0}$. The set of all such feature embeddings is given as M.

2.6 Mesh Feature Optimisation

To incorporate texture information into the tracking algorithm, we constrain any surface texture to have constant material point coordinates. Therefore, we now seek to find a surface parameterised with bi-cubic Hermite elements that closely matches the point cloud.

To take into account the detected feature points across all time points, we used the following error function:

$$\varepsilon_F = \sum_t^T \sum_f^F ||\mathbf{x}_{f,t} - \psi(\xi_f, \mathbf{P}_{s,t})||^2, \tag{4}$$

which assesses the total Euclidean distance between a feature's observed world position $\mathbf{x}_{f,t}$, and position modelled by the bi-cubic Hermite elements and material coordinate ξ_f. We include the material coordinates of all of the features as degrees of freedom. While this does introduce more gauge symmetries into an un-regularised optimisation [11], this is still fixed by the use of the Sobolov regulariser of Eq. 2.

This is added to Eq. 3 to give the SIFT augmented loss function:

$$\mathbb{P}_{s,t}, \mathbb{M}_f = \underset{\mathbf{P}_{s,t}, M_f}{\operatorname{argmin}} \left(\alpha \sum_t^T \sum_e^E S(e, \mathbf{P}_{s,t})^2 + \beta \sum_t^T \sum_e^E d(\psi(e, \mathbf{P}_{s,t}), C_t)^2 \right. \\ \left. + (1 - \alpha - \beta) \sum_t^T \sum_f^F \|\mathbf{x}_{ft} - \psi(\xi_f, \mathbf{P}_{s,t})\|^2 \right), \tag{5}$$

which now, via SIFT features, incorporates texture information as data terms that must be explained by the model. The weighting parameters used for the optimisation were $\alpha = \frac{1}{111}, \beta = \frac{10}{111}$.

2.7 Ground Truth Recovery

Marker Based Ground Truth. Adhesive backed ArUco markers were distributed evenly over the surface of the skin, but without following a predefined pattern [5]. Each marker "dictionary" provided contains up to a 1000 markers. The markers used in this test were 36, 37, 38, and 39.

These markers provided a direct measurement of their location for every frame within which they were visible. The markers were detected in each image frame and then optimised with the bundle adjustment framework to measure their world coordinates at each time.

To evaluate the performance of the markerless tracking methods in comparison to the ground truth, the mesh surface was used to predict the position of the marker at each time point. To evaluate a tracked mesh surface, the 3D motion of the ξ coordinate closest to the marker in the initial frame was compared to the bundle adjustment-based 3D motion estimate of the marker.

Registration Based Evaluation of Residual Error. For a given shape parameterisation $\mathbb{P}_s$, the parameterised deformation of the surface also predicts the 2D deformation field seen by each imaging camera. A dense set of points in the imaging plane were used as evaluation points, embedded into the mesh as in Sect. 2.5, and then the estimated motion observed by all cameras was generated using the change in the embedding function ψ from parameters $\mathbb{P}_{s,t} \rightarrow \mathbb{P}_{s,(t+1)}$. A dense pairwise comparison using phase-based registration [4] with a window size of 128 px × 128 px was used to the measure the residual motion (Fig. 3).

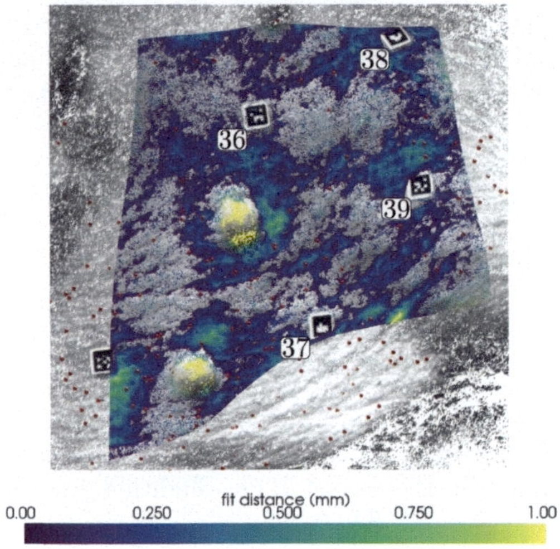

Fig. 3. The fitted surface, and the reconstructed point cloud, at time 0. The locations of detected surface features during time 0 are visible as small red spheres on the object surface. (Color figure online)

3 Results

The fitted surface matched the reconstructed surface well for all frames, but larger errors were visible around the nipples. The pig skin was highly textured, and had bumps on the order of 1 mm. The deformations recovered ranged up to 10 mm in magnitude. Both mesh based methods matched the shape deformation of the recovered point cloud (Fig. 4).

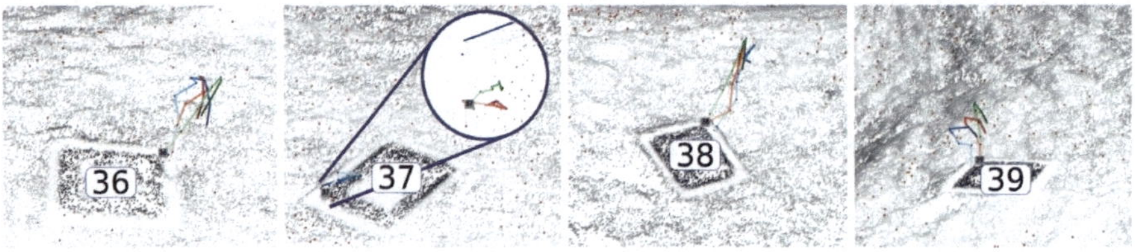

Fig. 4. The estimated deformation of the skin in the local region of each marker. The gold standard method is in green, shape based fitting in blue, and with SIFT augmented fitting in red. The colour scale of the vectors increases in intensity with time. (Color figure online)

We found that the improvement in tracking from using texture information depended heavily on the density of data surrounding each point. Markers 37 and 39 were comparatively well surrounded by embedded feature points, and had much lower errors than markers 36 and 38, which were in regions of the

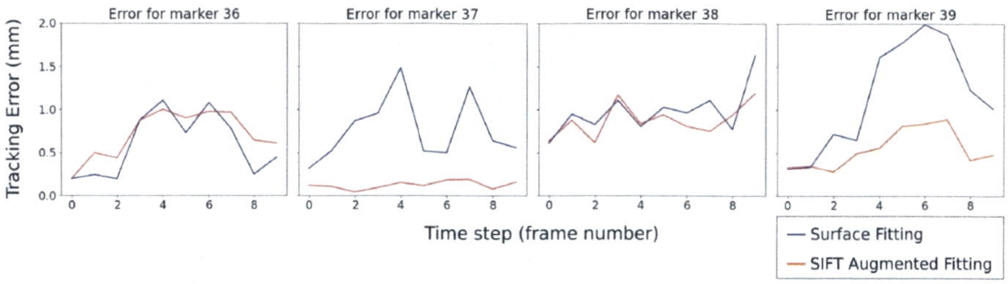

Fig. 5. The error in the recovered deformation, as compared to the ground truth. The SIFT augmented is in red, while the estimate from surface fitting is in blue. (Color figure online)

porcine skin that did not detect as many markers over the surface. The initial error varied from marker to marker, but was nonzero as the result of the residual fitting error over the surface.

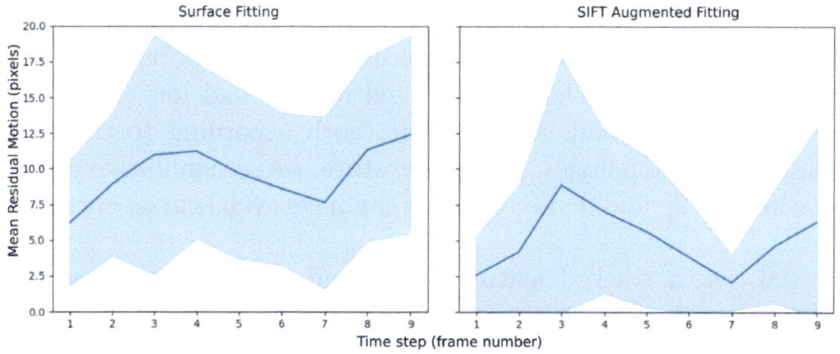

Fig. 6. The residual motion of the skin surface in each camera, found using pairwise registration [4], presented as mean (line) and standard deviation (shaded) over time. The surface fitting-based method had higher average error, greater than a standard deviation away from zero, and gradually increases over the deformation. In contrast, the average error for SIFT augmented fitting is smaller, and typically included a significant number of points with small errors - especially so at time points 1 and 7.

The registration results, Fig. 7, show that incorporating texture information into the fitting regime reduced the residual motion. Using texture information lowered the average error, with some measurements having zero error. This shows that the texture matching as a result of the SIFT embedding gives close agreement with texture matching using phase-based registration.

Comparing the unexplained motion in camera 0 between time 0 and time 1, the shape-based loss had up to 20 px of error over the whole of the surface. In the texture-based loss this is reduced to approximately 15 px maximum error, in regions with low numbers of detected SIFT features. In regions where dense SIFT features were found, the error was lower or close to zero. This matches he overall behaviour shown in Fig. 6.

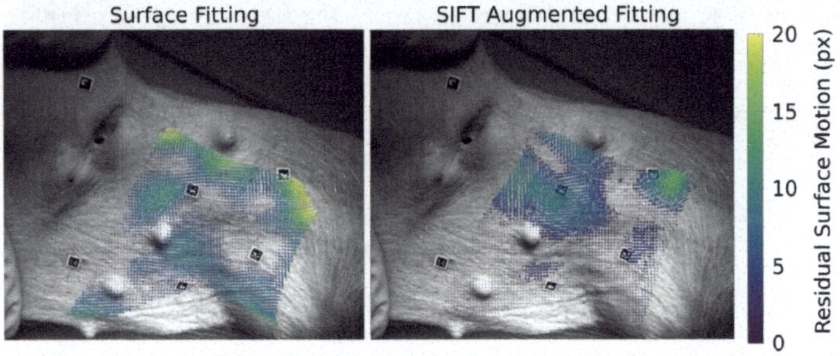

Fig. 7. The measured residual reprojection error as seen by an imaging camera at time 0 using pairwise registration [4]. Error presented is shown between time 0 and time 1.

4 Discussion

Our results show that, although incorporating texture information in the form of sparse features improves tracking accuracy, sparse features have weak influence over their surroundings. The local density of detections was inconsistent, despite similar lighting, focus, and the greater texture present in porcine skin due to the presence of large wrinkles. Densely detected regions had low error, but regions with few SIFT features had notable drift, both according to the marker and residual surface motion metrics. In the literature, we see similarly sparse features used for tracking [1,2], and dense methods can have even sparser correspondences [3].

In this dataset, a tracked feature was found approximately every $10\,\text{mm}^2$, resulting in an error of approximately 1 mm developing over the tracked four seconds. For more accurate tracking of the surface, these results suggest that some form of dense measurement over the surface is likely necessary, as even a greater number of cameras than is typically used in soft tissue tracking applications could not achieve sufficient point density over the surface.

5 Conclusion

We have demonstrated a SIFT based sparse surface tracker, which describes the deformation of skin as a bi-cubic Hermite mesh. However, we also showed that a surface described by sparse features can have significant drift, even when the motion of the sparse features is well explained by the surface model.

6 Future Work

In the future we aim to address the two possible mechanisms of error, and add dense tracking to the surface model, which will likely include the use of dense registration. Incorporating a dense estimate of deformation will provide more information, even when the feature detection methods produce no plausible features in a region.

References

1. Zhou, H., Jagadeesan, J.: Real-time surface deformation recovery from stereo videos. In: Shen, D., et al. (eds.) MICCAI 2019. LNCS, vol. 11764, pp. 339–347. Springer, Cham (2019). https://doi.org/10.1007/978-3-030-32239-7_38
2. Collins, T., Bartoli, A., Bourdel, N., Canis, M.: Robust, real-time, dense and deformable 3D organ tracking in laparoscopic videos. In: Ourselin, S., Joskowicz, L., Sabuncu, M.R., Unal, G., Wells, W. (eds.) MICCAI 2016. LNCS, vol. 9900, pp. 404–412. Springer, Cham (2016). https://doi.org/10.1007/978-3-319-46720-7_47
3. Richey, W.L., Heiselman, J.S., Ringel, M.J., Meszoely, I.M., Miga, M.I.: Soft tissue monitoring of the surgical field: detection and tracking of breast surface deformations. IEEE Trans. Biomed. Eng. **70**(7), 2002–2012 (2023). https://doi.org/10.1109/TBME.2022.3233909
4. HajiRassouliha, A., Tang, E.J.L.P., Taberner, A.J., Nash, M.P., Nielsen, P.M.: A method for three-dimensional measurements using widely angled stereoscopic cameras. In: 2019 IEEE International Instrumentation and Measurement Technology Conference (I2MTC), pp. 1–5 (2019). https://doi.org/10.1109/I2MTC.2019.8827086
5. Garrido-Jurado, S., Muñoz-Salinas, R., Madrid-Cuevas, F.J., Marín-Jiménez, M.J.: Automatic generation and detection of highly reliable fiducial markers under occlusion. Pattern Recogn. **47**(6), 2280–2292 (2014). https://doi.org/10.1016/j.patcog.2014.01.005
6. Zerpa, J.M.P., Talou, G.D.M., Blanco, P.J.: A new robust formulation for optical-flow/material identification problems. Comput. Methods Appl. Mech. Eng. **351**, 766–788 (2019). https://doi.org/10.1016/j.cma.2019.04.012
7. Otero, I.R., Delbracio, M.: Anatomy of the SIFT method. Image Process. Line **4**, 370–396 (2014). https://doi.org/10.5201/ipol.2014.82
8. Tyszkiewicz, M.J., Fua, P., Trulls, E.: DISK: learning local features with policy gradient. arXiv preprint arXiv:2006.13566 (2020)
9. Lindenberger, P., Sarlin, P.-E., Pollefeys, M.: LightGlue: local feature matching at light speed. arXiv preprint arXiv:2306.13643 (2023)
10. Brown, D.C.: Decentering distortion of lenses. In: Photogrammetric Engineering and Remote Sensing (1966)
11. Triggs, B., McLauchlan, P.F., Hartley, R.I., Fitzgibbon, A.W.: Bundle adjustment — a modern synthesis. In: Triggs, B., Zisserman, A., Szeliski, R. (eds.) IWVA 1999. LNCS, vol. 1883, pp. 298–372. Springer, Heidelberg (2000). https://doi.org/10.1007/3-540-44480-7_21
12. Hartley, R.I., Zisserman, A.: Multiple View Geometry in Computer Vision, 2nd ed. Cambridge University Press (2004)
13. Wang, Y., Zeng, Z., Guan, T., et al.: Adaptive patch deformation for textureless resilient multi-view stereo. In: The Proceedings of the IEEE/CVF Conference on Computer Vision and Pattern Recognition, pp. 1621–1630 (2023)
14. Fernandez, J.W., Mithraratne, P., Thrupp, S.F., Tawhai, M.H., Hunter, P.J.: Anatomically based geometric modelling of the musculo-skeletal system and other organs. Biomech. Model. Mechanobiol. **2**(3), 139–155 (2004). https://doi.org/10.1007/s10237-003-0036-1

Towards Determining Mechanical Properties of Brain-Skull Interface Under Tension and Compression

Sajjad Arzemanzadeh[1(✉)], Benjamin F. Zwick[1], Karol Miller[1], Tim Rosenow[2], Stuart I. Hodgetts[3,4], and Adam Wittek[1]

[1] Intelligent Systems for Medicine Laboratory (ISML), School of Engineering, The University of Western Australia, Perth, WA 6009, Australia
sajjad.arzemanzadeh@research.uwa.edu.au

[2] Western Australia National Imaging Facility, The University of Western Australia, Perth, WA 6009, Australia

[3] Spinal Cord Repair Laboratory, School of Human Sciences, The University of Western Australia, Perth, WA 6009, Australia

[4] Perron Institute for Neurological and Translational Science, Verdun Street, Nedlands, WA 6009, Australia

Abstract. Computational biomechanics models of the brain have become an important tool for investigating the brain responses to mechanical loads. The geometry, loading conditions, and constitutive properties of such brain models are well-studied and generally accepted. However, there is a lack of experimental evidence to support models of the layers of tissues (meninges) connecting the brain to the skull (brain-skull interface), which determine boundary conditions for the brain. We present a new protocol for determining the biomechanical properties of the brain-skull interface and present the preliminary results (for a small number of tissue samples extracted from sheep cadaver heads) obtained using this protocol. The method consists of biomechanical experiments using brain tissue and brain-skull complex (consisting of the brain tissue, brain-skull interface, and skull bone) and comprehensive computer simulation of the experiments using the finite element (FE) method. Application of the FE simulations allowed us to abandon the traditionally used approaches that rely on analytical formulations that assume cuboidal (or cylindrical) sample geometry when determining the parameters that describe the biomechanical behaviour of the brain tissue and brain-skull interface. In the simulations, we used accurate 3D geometry of the samples obtained from magnetic resonance images (MRIs). Our results indicate that the behaviour of the brain-skull interface under compressive loading appreciably differs from that under tension. Rupture of the interface was clearly visible for tensile load while no obvious indication of mechanical failure was observed under compression. These results suggest that assuming a rigid connection or frictionless sliding contact between the brain tissue and skull bone, approaches often used in computational biomechanics model of the brain, may not accurately represent the mechanical behaviour of the brain-skull interface.

Keywords: Brain-skull interface · Meninges · Biomechanical properties · Finite element simulation · Biomechanical experiments

M. Kobielarz et al. (Eds.): MICCAI 2024, LNBE, pp. 114–128, 2025.
https://doi.org/10.1007/978-3-031-94128-3_11

1 Introduction

Computational biomechanics head models implemented using the finite element (FE) method were first reported in the early 1980s [1], and since then, have been a subject of ongoing research effort. Nowadays, comprehensive brain models encompass a wide variety of applications including crash injury prediction and computer-assisted neurosurgery [2–5]. Computational biomechanics brain models include four main elements: (i) the geometry of the brain, typically obtained from electronic brain atlases or magnetic resonance images (MRIs); (ii) boundary and contact conditions; (iii) loading conditions, and (iv) the material laws and material properties of intracranial components. The brain geometry, loading conditions and constitutive properties have been extensively studied. However, there is a lack of experimental evidence to support the currently used models of the brain-skull interface (layers of tissue located between the brain cortical surface and skull inner surface) that determines boundary conditions for the brain [6]. Therefore, parametric studies have been conducted to evaluate the effects of different approaches for modelling the brain-skull interface on the brain responses predicted by computational biomechanics models [6–8].

The brain-skull interface is composed of three main membranous layers (meninges): (i) Dura mater (ii) Arachnoid and (iii) Pia mater [9]. These layers are sandwiched between the cranial inner surface and the brain's outer surface (Fig. 1a). Space between the arachnoid and pia mater (the subarachnoidal space) is filled with cerebrospinal fluid (CSF) and arachnoid trabeculae that are extended from the inner surface of the arachnoid to the pia mater [10, 11]. The exact anatomical structure and mechanical properties of these sublayers are still a matter of debate as scarce quantitative information exists regarding their mechanical properties and interactions [6, 9].

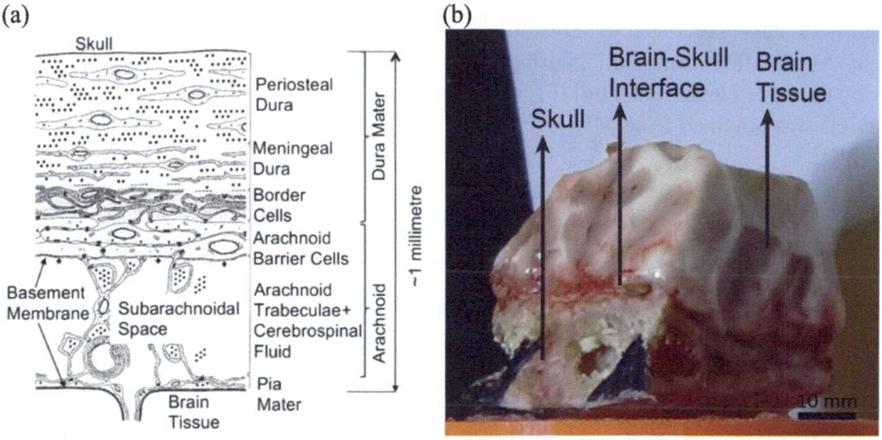

Fig. 1. (a) Anatomical structure of the brain-skull interface (meninges). Copied from [6]. (b) Brain-skull complex sample extracted from sheep cadaver head showing the brain tissue, brain-skull interface, and skull bone.

One possible method of investigating the mechanical behaviour of the brain-skull interface involves conducting mechanical tests using samples of tissues that form the brain-skull interface extracted from animal cadaver heads. For instance, Jin et al. [12] investigated the mechanical properties of the pia-arachnoid complex extracted from a cadaver bovine head under in-plane tension, normal traction, and shear loadings. They found that the pia-arachnoid complex exhibits nonlinear viscoelastic material behaviour and is significantly stiffer under in-plane tension than under normal traction and shear. Although studies of meninges and pia-arachnoid tissue samples provide valuable insight into the biomechanical behaviour of the anatomical components of the brain-skull interface, application of their results in comprehensive biomechanical simulations of the brain poses significant challenges due to the risk of inducing damage in thin tissue layers when excising the samples and complexity that would be required to construct the brain-skull interface model that explicitly represents anatomical components of the brain-skull interface and mechanical interactions between these components.

These challenges can be overcome by conducting experiments on samples of the brain-skull complex, which includes skull bone, brain-skull interface, and brain tissues (Fig. 1b). This approach allows the brain-skull interface to remain intact when excising the samples. In 2015, Agrawal et al. [13] conducted compressive tests on the brain-skull complex samples extracted from sheep cadaver heads. They used a finite element model consisting of skull bone and brain tissue with frictionless contact between them to analyse their experiments. Although they used a contact digital scanner to obtain accurate 3D geometry of the skull bone for the model, simplified cuboidal geometry of the brain tissue sample was assumed.

In this study, we demonstrate a new method for investigation of the mechanical properties of the brain-skull interface. The method consists of uniaxial tension and compression experiments on the brain tissue and brain-skull complex samples, with computational biomechanics modelling applied to analyse the experimental results to determine the subject-specific constitutive properties of the brain tissue and mechanical behaviour of the brain-skull interface. Modelling of the experiments was conducted using the finite element method (FEM) and accurate 3D geometry of the samples for the models was obtained from MRIs.

2 Methods

2.1 Sample Preparation

The experiments, including sample preparation, MRI acquisition, and biomechanical tests, were conducted at the laboratory facilities of the Harry Perkins Institute of Medical Research (Nedlands, WA, Australia). Two sheep cadaver skinned heads were collected from an abattoir (Dardanup Butchering Company) and transported to the laboratory (with The University of Western Australia Biosafety Approval F69199). As the brain-skull complex is a structure (consisting of the brain tissue, brain-skull interface, and skull bone), the subject-specific material properties of brain tissue first needed to be determined. Therefore, biomechanical tests were conducted on two sets of cuboidal samples extracted from each head: (i) brain tissue and (ii) a brain-skull complex consisting of the brain tissue, meninges, and skull bone. Experiments were completed within 24 h after the animals' death. This timeframe is shorter than the delay duration that could pose a substantial risk of alterations of the brain tissue characteristics [14].

Figure 2a and 2b show the sheep cadaver skinned head before and after the sample extraction. To avoid damage to the brain-skull interface, a dental technician drill set (Saeshin Strong 206/H450 by Saeshin, Daegu, Korea) equipped with a Dynex cutting disc (22x0.3 mm) by Renfert (www.renfert.com) was used to cut the skull. Subsequently, a microtome blade (Erma Patho Cutter HP-R 35/75mm) was used to cut the brain tissue in a dorsal-ventral direction (in sagittal and coronal planes). Figure 2c and 2d show the cutting disc and microtome blade, respectively. The cutting lines and the locations of the three extracted parts are shown in Fig. 2a. To minimise the risk of heat- and mechanically induced damage to the brain tissue and brain-skull interface, dissection of skull bone was conducted in two stages:

(1) An initial cut that penetrated through most of the bone thickness.
(2) A final fine cut at a slow feed speed to prevent the discs from cutting through the brain tissue.

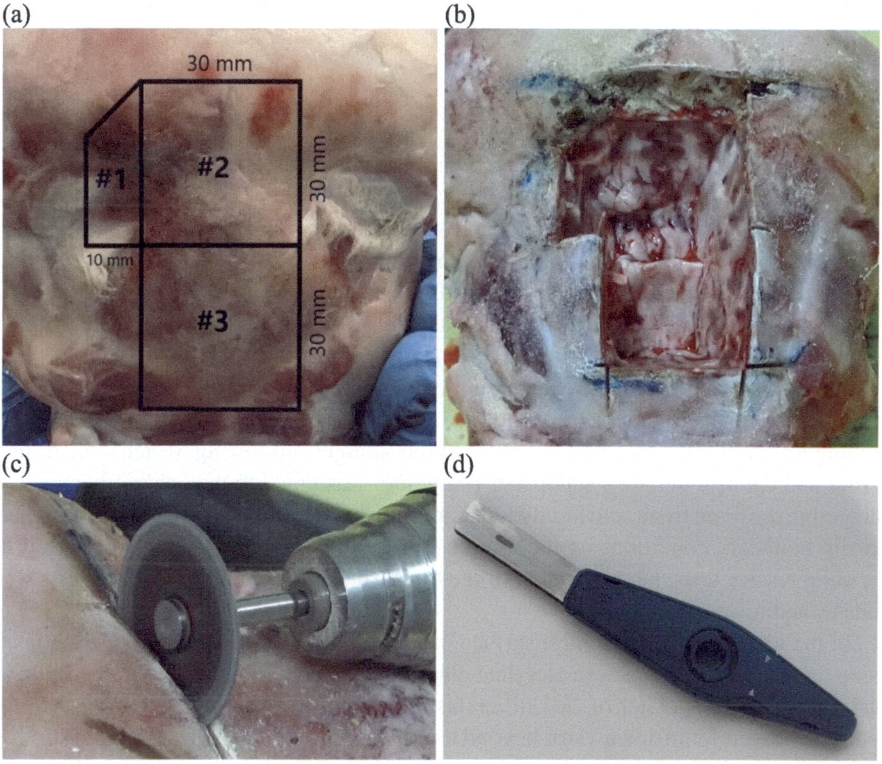

Fig. 2. Cadaver sheep head (a) before and (b) after sample extraction. The cutting lines and part numbers are illustrated in (a). (c) Dental technician drill equipped with cutting disc. (d) Microtome blade.

Part 1 was removed first to allow access to the area beneath Part 2 and to cut it horizontally (in the transverse plane). Part 2 was used as the brain-skull complex sample. After the extraction of Part 2, the area beneath Part 3 became accessible. Part 3 was used as the brain tissue sample following the removal of the skull bone. In total, one brain tissue sample (intended dimensions of around 27 mm x 27 mm x 17 mm) and one brain-skull complex sample (intended brain tissue dimensions of around 26 mm x 27 mm x 17 mm) were extracted from each cadaver sheep head. The T2-weighted MRIs of all the samples were then acquired. The samples were not frozen at any time and were kept hydrated inside containers filled with 100 g of 5% saline solution.

2.2 Acquisition of Sample Images

Magnetic resonance images (MRIs) of the sheep cadaver heads and tissue samples were acquired using Bruker Biospec 9.4 T animal scanner at the Western Australia node of the National Imaging Facility (NIF) located at the Harry Perkins Institute of Medical Research (Nedlands, WA, Australia). We used a 3D T2-weighted TurboRARE sequence (repetition time = 2200 ms, echo time = 34 ms, RARE factor = 12, partial Fourier acceleration factor = 1.4, matrix size 396x192x48), resulting in a final resolution of 0.25 mm x 0.25 mm x 0.5 mm.

2.3 Experiments

The UniVert (CellScale, www.cellscale.com), a specialised portable biomaterial testing system driven by a stepper motor under closed loop control, was used for uniaxial compression and tensile tests. The testing system was equipped with a 10 N LFT-25-1KG load-cell (accuracy of 0.2% of the maximum load, which implies 0.02 N) and displacement transducer. The force and loading head displacements were recorded with a sampling rate of 100 Hz. The experimental protocol was based on Miller [15, 16] and Miller and Chinzei [17, 18]. Dorsal (cortical) surface of the brain tissue samples and outer skull surface for the brain-skull complex samples were glued to the UniVert testing system platen using a fast-curing cyanoacrylate glue (Fix&Go Supaglu by Selleys'). The loading was applied to the ventral surface of the samples. Following Miller [15, 16] and Miller and Chinzei [17, 18], sandpaper (80 grade) was glued to the loading head when subjecting the brain tissue and brain-skull complex samples to compression. This implied no-slip boundary conditions on the sample surface in contact with the loading head in the compressive tests, i.e. movement of the top surface of the samples in such tests was constrained to vertical direction.

In the tensile tests, the ventral surface of the sample was glued to the loading head using the same fast-curing cyanoacrylate glue that was used to attach the sample's dorsal surface to the bottom platen of the UniVert testing system. To ensure adhesion, the loading head was moved 1 mm downwards beyond its initial contact point with the ventral surface of the samples. It was held in this position for 20 s, then returned to the original contact point and held there for additional 60 s to provide stress relaxation before the start of the tensile tests. This procedure follows Miller [16].

In all experiments conducted in this study, the loading speed was 0.3 mm/s which implies a nominal strain rate of approximately 0.02/s.

All tests were conducted at room temperature.

The experiments were recorded using two digital cameras: (1) High resolution DFK 33UX264 scientific camera (resolution of 2048 x 2048, 15 Frames/s) that provided the frontal view of the test samples; and (2) Sony HDR-CX625 camcorder (resolution of 1920 x 1080, 25 Frames/s) that provided the general oblique view of the test samples.

The brain tissue and brain-skull complex samples extracted from the first sheep cadaver head were subjected to tension and referred to as H1B-T and H1S-T, respectively. The samples extracted from the second sheep cadaver head were subjected to compression and referred to as H2B-C and H2S-C, respectively. Figure 3 shows the samples in initial (unloaded) configuration before starting the mechanical tests.

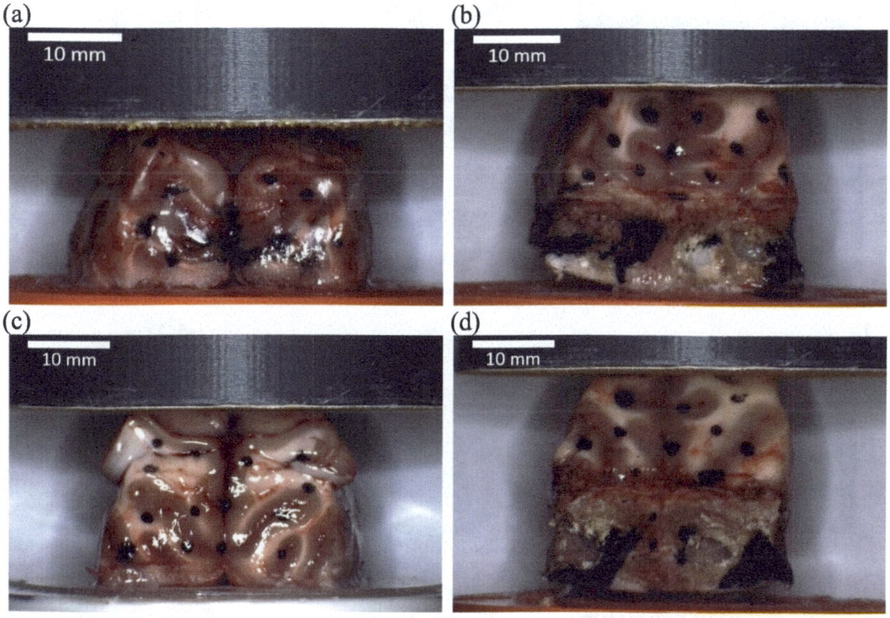

Fig. 3. The samples before starting the tests. (a) brain tissue sample used in tensile test (H1B-T), (b) brain-skull complex used in tensile test (H1S-T), (c) brain tissue sample used in compressive test (H2B-C), and (d) brain-skull complex used in compressive test (H2S-C).

2.4 Modelling

2.4.1 Finite Element Mesh Generation

3D Slicer software [19] was used to semi-automatically segment the MRIs of the samples to extract their accurate 3D geometry. 3D surfaces constructed from the MRI segmentations using 3D Slicer software were used as a starting point to create finite element meshes of the samples. Meshing was done using Coreform Cubit 2024.3 (Coreform, www.coreform.com/products/coreform-cubit) finite element mesh generator and imported into

Abaqus 2023 finite element software (Dassault Systèmes, www.3ds.com/products/simulia/abaqus). This allowed us to create high-quality (mean Jacobian between 0.62 and 0.75) fully hexahedral (using 8-noded elements) meshes of the samples (Fig. 4 and Table 1).

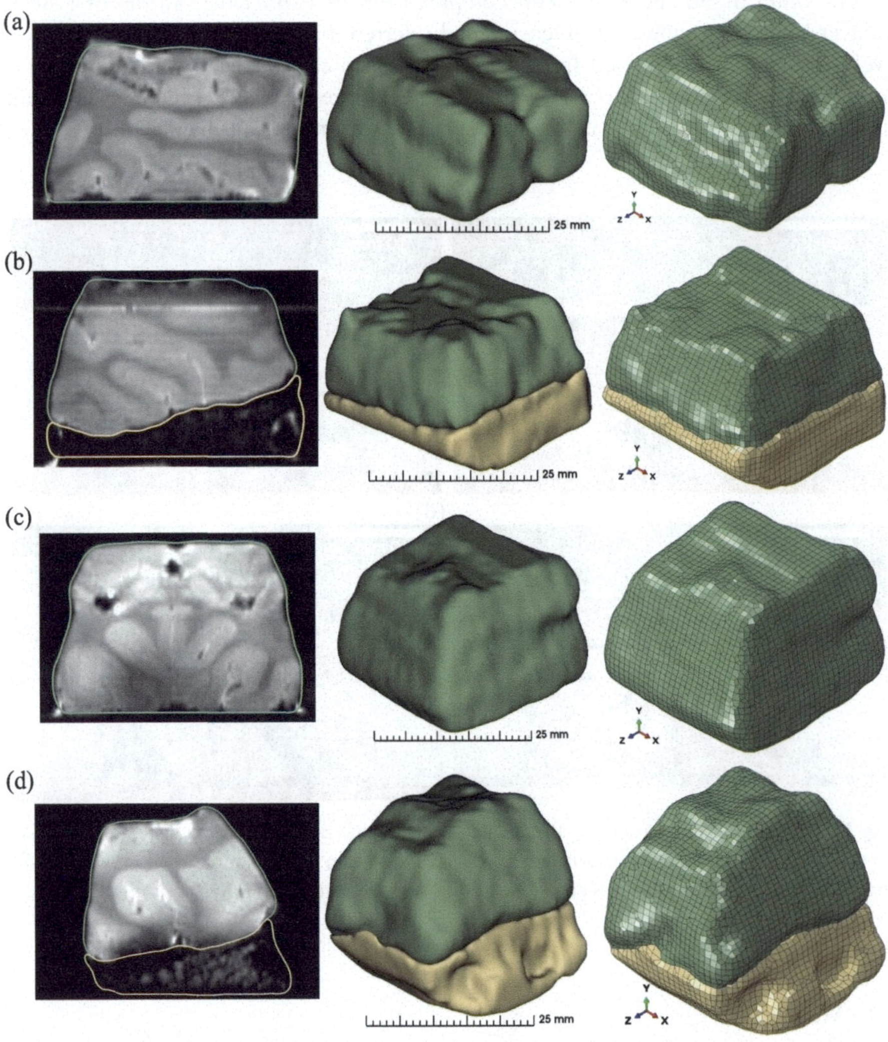

Fig. 4. The MRI of the samples with highlighted boundary of material segmentation (left column), 3D representation of the segmentation (middle column), and meshed geometries imported into Abaqus (right column). Brain: green, and skull: yellow. (a) brain tissue sample used in tensile test (H1B-T), (b) brain-skull complex used in tensile test (H1S-T), (c) brain tissue sample used in compressive test (H2B-C), and (d) brain-skull complex used in compressive test (H2S-C).

Table 1. The finite element mesh metrics of the brain tissue in samples.

Sample	Number of nodes	Number of elements	Mean Jacobian
H1B-T	16938	14969	0.6745
H1S-T	15212	13376	0.6246
H2B-C	21903	19658	0.7504
H2S-C	16657	14699	0.6815

2.4.2 Finite Element Solver and Element Formulation

All simulations were conducted using the Abaqus explicit dynamics non-linear finite element solver. We used stiffness-based hourglass control and automated time stepping. To prevent volumetric locking, we used under-integrated hexahedral element formulation (C3D8R element type in Abaqus).

2.4.3 Boundary Conditions and Loading

In all tests of the brain tissue samples (tension: sample H1B-T; compression: sample H2B-C), the bottom (dorsal) surfaces of the samples were glued to the UniVert testing system platen. Hence, in the FE models of these tests, the bottom nodes of the samples were rigidly constrained. In all tests of the brain-skull complex samples (tension: sample H1S-T; compression: sample H2S-C), the outer surface of the skull bones was glued to the UniVert testing system platen. Hence, in the FE models of such tests, the skull bones were rigidly constrained.

In the tensile tests, the top surface of the sample was bonded to the loading head using fast-curing cyanoacrylate glue. In the compressive tests, the sandpaper glued to the loading head constrained movement of the sample surface in contact with the loading head to vertical direction. Therefore, when modelling the tensile and compressive tests, loading was defined by prescribing the displacement (rate of 0.3 mm/s) in the vertical direction on the model nodes representing the top surfaces of the samples. The displacement was applied using a smooth step procedure in the Abaqus finite element software.

In the FE models of the brain-skull complex samples (tension: sample H1S-T; compression: sample H2S-C), the brain-skull interface was modelled as a rigid tie between the brain tissue and skull. This approach has been used in several previous studies [6, 20].

2.4.4 Constitutive Models and Parameters

The skull was modelled as rigid since it is orders of magnitude stiffer than the brain tissue. Following Miller [15, 16] and Miller and Chinzei [17, 18], we used a first-order Ogden-type hyperelastic model to describe the material behaviour of the brain tissue:

$$W(\lambda_1, \lambda_2, \lambda_3) = \frac{2\mu}{\alpha^2}\left(\lambda_1^\alpha + \lambda_2^\alpha + \lambda_3^\alpha - 3\right) + \frac{1}{D}(J_{el} - 1)^2, \tag{1}$$

where λ_1, λ_2 and λ_3 are the principal stretches, J_{el} is the third strain invariant (also referred to as the volumetric strain), μ is the shear modulus, α is the dimensionless constant, and D determines the compressibility of the material:

$$D = \frac{2}{K}, \tag{2}$$

where K is the bulk modulus. The relationship between the bulk modulus, shear modulus, and Poisson's ratio ν is:

$$\nu = \frac{3\frac{K}{\mu} - 2}{6\frac{K}{\mu} + 2}. \tag{3}$$

As there is consensus in the literature that the brain and other soft tissues are nearly incompressible [21], we used Poisson's ratio of 0.49 for the brain tissue. When determining the subject-specific brain material constants μ and α of the brain tissue, the force-displacement curves from the FE models were compared to those obtained in the experiments. We used an iterative process in which the constants were calibrated to match the force-displacement curves observed in the experiments. Following Miller [17], the calibration was done for uniaxial strain between –0.3 (compression) and 0.2 (tension). This resulted in a unique set of material constants that accurately describe the brain tissue material behaviour under both tension and compression.

The subject-specific material constants were then used when simulating the experiments on the brain-skull complex samples extracted from the same sheep cadaver head to investigate the biofidelity of modelling the brain-skull interface as a rigid tie between the brain tissue and skull.

3 Results

Figure 5 shows the force-displacement curves obtained in the experiments and from the FE models of the brain tissue samples subjected to tension (sample H1B-T) and compression (sample H2B-C). The maximum difference between the force obtained from the models and force measured experimentally was 0.04 N (10% of the experimentally measured force) for tensions and 0.15 N (11% of the experimentally measured force) for compression. These values indicate a very close agreement between the modelling and experimental results. Table 2 reports the subject-specific Ogden hyperelastic material constants (shear modulus μ and dimensionless parameter α, see Eq. 1) of the brain tissue determined in this study. The dimensionless parameter α we obtained here is very close to that reported by Agrawal et al. [13] for sheep brain tissue. The results reported in Fig. 5 confirm the previous findings [17] that the brain tissue stiffness under tension is appreciably lower than under compression.

Figure 6 shows the force-displacement curves obtained in the experiments and from the FE models of brain-skull complex samples (H1S-T and H2S-C) subjected to compression and tension. Figure 7 shows these samples in their deformed states under loading. In the tensile test of the brain-skull complex sample (H1S-T), the force continuously increased for the displacement of up to 3.3 mm (see Fig. 6a). It rapidly dropped at

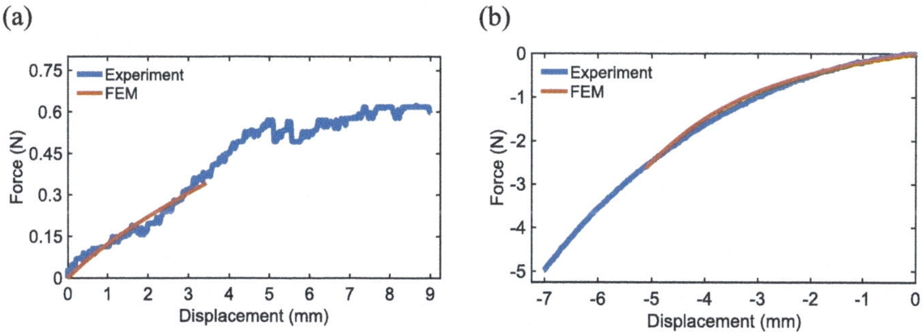

Fig. 5. Results for the brain tissue samples. Comparison of the force-displacement curves observed in the experiments and obtained from the FE models when analysing the experimental results to determine the brain tissue material constants for (a) the brain tissue sample subjected to tension (H1B-T); and (b) the brain tissue sample subjected to compression (H2B-C). The term displacement refers to the displacement of the test system loading head. Negative displacement means compression; and positive displacement means tension (sample elongation).

Table 2. Ogden hyperelastic material constants of the brain tissue determined in this study through application of the FE models to analyse the experimental results. The constants were determined from the compressive and tensile tests of the brain tissue samples (see Fig. 3a and c) for nominal strain between -0.3 (compression) and 0.2 (tension).

Material	Loading speed (mm/s)	μ (Pa)	α
Brain tissue	0.3	1200	–6.3

3.3 mm and then started to increase again. We interpret this force drop as an indication of the brain-skull interface mechanical failure and the start of detachment of the brain-skull interface from the skull bone. This detachment initially occurred only locally (see Fig. 7b). Then, it rapidly expanded until the entire interface detached from the skull, with the connection between the brain and skull maintained solely by the bridging veins (see Fig. 7c and d). We attribute the continuous increase in the force magnitude following the sudden drop at 3.3 mm to elongation of the bridging veins and dura mater connected to falx. Elongated bridging veins are clearly visible in Fig. 7d.

As inspection of the skull bone forming the brain-skull complex samples after the experiments indicated that the dura remained intact and firmly attached to the skull (see Fig. 8a), we suggest that the failure of the brain-skull interface may have occurred in the layers spanning arachnoid, arachnoid barrier cells, and border cells (see Fig. 8b).

Modelling of the tensile test of the brain-skull complex (H1S-T sample) using the assumption that the brain is rigidly attached to the skull resulted in a force-displacement relationship that accurately represents the experimental results for the sample elongation of up to 2 mm. This suggests that the tensile stiffness of the brain-skull interface, including the bridging veins, dura, and falx, may be appreciably higher than that of the brain tissue. This interpretation aligns with previous findings [22–24].

The key shortcoming of modelling the brain-skull interface as a rigid connection (tie) between the brain and skull is that it does not account for the interface failure observed in our experiments.

The force-displacement relationship of the brain-skull sample complex under tension was nearly linear for the displacement between 2 mm and 3 mm (Fig. 6a). One possible explanation for this behaviour may be the contribution of forces induced in the bridging veins and falx. The effects of the bridging veins and falx may also be a possible explanation for the relatively large difference (0.23 N, 30% of the experimentally measured force) between the experimental and modelling results for the brain-skull sample under tension (sample H1S-T).

For compression, the force-displacement curve of the brain-skull complex sample (H2S-C) (Fig. 6b) exhibited a similar trend to that of the brain tissue sample (H2B-C) (Fig. 5b). For the sample top surface displacement of up to around 4 mm (nominal compressive strain of around -0.23), the modelling results agreed well with the experiments. For the displacement exceeding 4 mm compression, the model tended to overestimate the forces measured in the experiments. For 5.1 mm compression, the difference the experimental and modelling results was 0.59 N (20% of the experimentally measured force). It may be tempting to explain this discrepancy by possible damage of the brain-skull interface and/or brain tissue for the compressive displacement of the sample top surface exceeding 4 mm. However, this interpretation of the results reported in Fig. 6b should be treated very carefully as no obvious indication of such damage under compressive loading was visually observed here (see Fig. 7e and f).

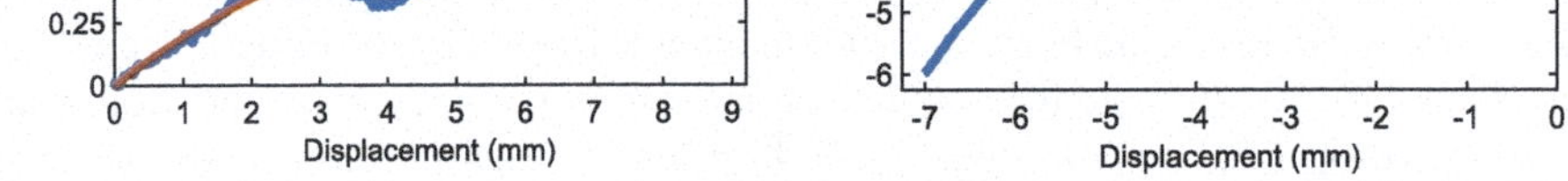

Fig. 6. Force-displacement curves for the samples of the brain–skull complex observed in the experiments and obtained from the FE models of the experiments. (a) Brain-skull complex sample subjected to tension (H1S-T); and (b) Brain-skull complex sample subjected to compression (H2S-C). In FE models, the brain tissue was rigidly attached to the skull bone. Positive displacement means tension (sample elongation) and negative displacement means compression.

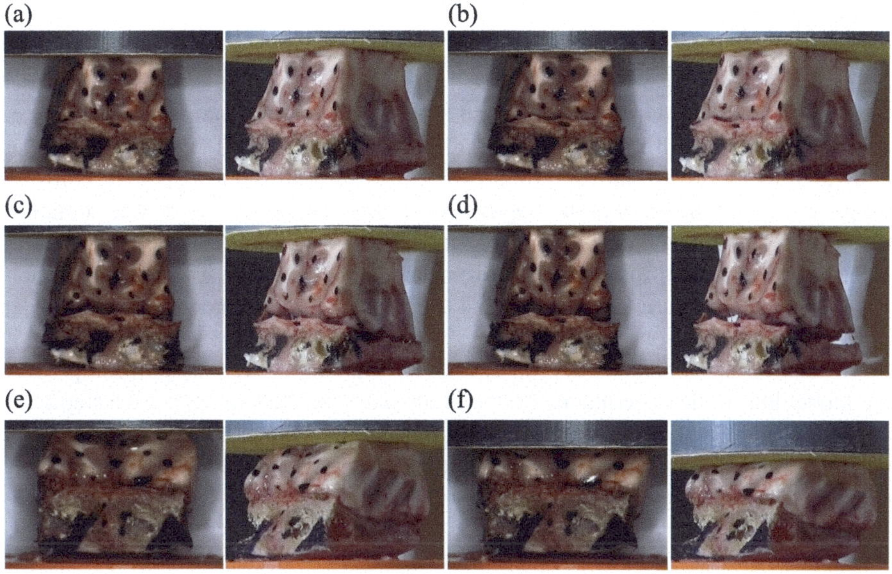

Fig. 7. Front and oblique views of the brain-skull complex samples in the experiments conducted in this study. (a) 3.3 mm elongation; (b) 3.6 mm elongation; (c) 6 mm elongation; (b) 9 mm elongation; (e) 3 mm compression; and (f) 6 mm compression. The load (displacement) was applied to the top (ventral) surfaces of the samples. Note detachment of the brain tissue from the skull for 6 mm elongation (c) and 9 mm elongation (d). The elongated bridging veins are indicated in (d).

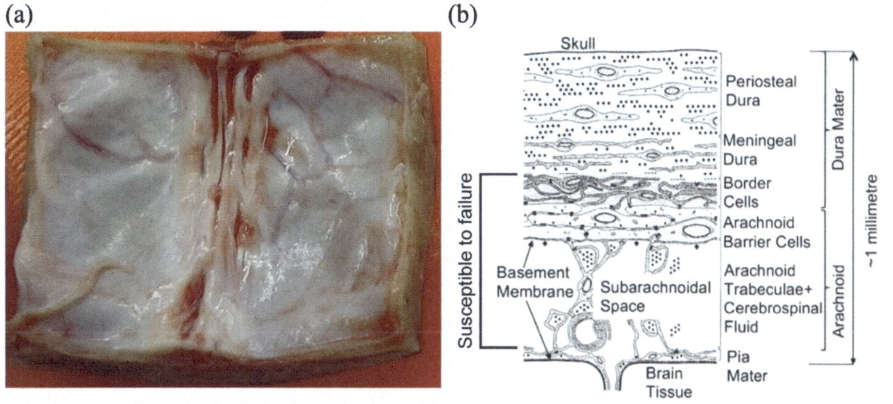

Fig. 8. (a) The dura remained intact and firmly attached to the skull after failure of the brain-skull interface under tension. (b) Anatomical structure of the brain-skull interface (meninges) with an indication of tissue layers that may be susceptible layers to failure under tensile load. Modified from [6].

4 Discussion

In this study, we investigated the mechanical properties of the brain-skull interface under compression and tension. Brain tissue and brain-skull complex samples (consisting of brain tissue, brain-skull interface, and skull bone) were extracted from sheep cadaver heads and subjected to uniaxial compression and tension. Subject-specific constitutive parameters of the brain tissue (Ogden constitutive model was used) were obtained through comprehensive finite element (FE) modelling of the experiments using the sample geometry obtained from magnetic resonance images (MRIs). The results are consistent with the literature and indicate that the brain tissue is appreciably stiffer under compression than tension.

The subject-specific material properties of the brain tissue were used in comprehensive modelling of the experiments on the brain-skull complex samples extracted from the same head to analyse the biomechanical behaviour of the brain-skull interface. We found that rigidly connecting the brain to the skull, one of the approaches for modelling boundary conditions for the brain used in the literature, may not be adequate as it does not capture the brain-skull interface failure under tension observed in the experiments conducted in the study. Our experiments also highlight the limitations of modelling the interactions between the brain and skull using a frictionless sliding contact, an approach used in our previous studies on neurosurgery simulation [3, 25–28], as it allows free separation of the brain cortical surface from the skull.

In this study, we focused on analysing the overall behaviour of the brain-skull interface. While we acknowledge the importance of cerebrospinal fluid (CSF) in influencing this behaviour, quantifying the effects of CSF would require further studies.

It can be concluded that this study confirms the feasibility of the proposed method for determining the biomechanical behaviour of the brain-skull interface using uniaxial compression and tension experiments combined with comprehensive modelling of the experiments and provides preliminary insight into such behaviour. However, it also highlights the need for conducting experiments on a larger number of brain tissue and brain-skull complex samples and indicates that an accurate representation of the biomechanical behaviour of the brain-skull interface would require models more advanced than those currently used in the literature. In particular, accounting for the mechanical failure of the brain-skull interface under tension may be one of the requirements such models need to satisfy.

Acknowledgement. This research was supported by the Australian Government through the Australian Research Council's ARC Discovery Projects funding scheme (project DP230100949). Author S. A. acknowledges support of the University of Western Australia Scholarship for International Research Fees and University of Western Australia (UWA) – Intelligent Systems for Medicine Laboratory (ISML) Higher Degree by Research Scholarship in Computational Biomechanics (partly funded from Australian Research Council's ARC Discovery Project DP230100949). The authors acknowledge the facilities and scientific and technical assistance of the National Imaging Facility, a National Collaborative Research Infrastructure Strategy (NCRIS) capability, at the Centre for Microscopy, Characterisation and Analysis, The University of Western Australia. The authors thank Mr Bernard Panizza of Dardanup Butchering Company (Bunbury, Western Australia, Australia) for help in obtaining sheep cadaver heads for the experiments conducted in this study.

References

1. Shugar, T.A., Katona, M.G.: Development of finite element head injury model. J .Eng. Mech. Div. **101**(3), 223–239 (1975)
2. Matsuda, T., et al.: Development of a human body model (THUMS Version 7) to simulate kinematics and injuries of reclined occupants in frontal collisions. In: 27th International Technical Conference on the Enhanced Safety of Vehicles (ESV) National Highway Traffic Safety Administration (2023)
3. Miller, K., et al.: Biomechanical modeling and computer simulation of the brain during neurosurgery. Int. J. Numer. Methods Biomed. Eng. **35**(10), e3250 (2019)
4. Carmo, G.P., et al.: Development, validation and a case study: the female finite element head model (FeFEHM). Comput. Methods Programs Biomed. **231**, 107430 (2023)
5. Lu, Y.C., et al.: A 3D computational head model under dynamic head rotation and head extension validated using live human brain data, including the falx and the tentorium. Ann. Biomed. Eng. **47**, 1923–1940 (2019)
6. Wang, F., et al.: Prediction of brain deformations and risk of traumatic brain injury due to closed-head impact: quantitative analysis of the effects of boundary conditions and brain tissue constitutive model. Biomech. Model. Mechanobiol. **17**, 1165–1185 (2018)
7. Yang, S., et al.: Assessment of brain injury characterization and influence of modeling approaches. Sci. Rep. **12**(1), 13597 (2022)
8. Zhou, Z., Li, X., Kleiven, S.: Evaluation of brain-skull interface modelling approaches on the prediction of acute subdural hematoma in the elderly. J. Biomech. **105**, 109787 (2020)
9. Haines, D.E., Harkey, H.L., Al-Mefty, O.: The "subdural" space: a new look at an outdated concept. Neurosurgery **32**(1), 111–120 (1993)
10. Walsh, D.R., et al.: Mechanical properties of the cranial meninges: a systematic review. J. Neurotrauma **38**(13), 1748–1761 (2021)
11. Mack, J., Squier, W., Eastman, J.T.: Anatomy and development of the meninges: implications for subdural collections and CSF circulation. Pediatr. Radiol. **39**, 200–210 (2009)
12. Jin, X., et al.: Constitutive modeling of pia–arachnoid complex. Ann. Biomed. Eng. **42**, 812–821 (2014)
13. Agrawal, S., Wittek, A., Joldes, G., Bunt, S., Miller, K.: Mechanical properties of brain–skull interface in compression. In: Doyle, B., Miller, K., Wittek, A., Nielsen, P. (eds.) Computational Biomechanics for Medicine. Springer, Cham (2015). https://doi.org/10.1007/978-3-319-15503-6_8
14. Rashid, B., Destrade, M., Gilchrist, M.D.: Influence of preservation temperature on the measured mechanical properties of brain tissue. J. Biomech. **46**(7), 1276–1281 (2013)
15. Miller, K.: Method of testing very soft biological tissues in compression. J. Biomech. **38**(1), 153–158 (2005)
16. Miller, K.: How to test very soft biological tissues in extension? J. Biomech. **34**(5), 651–657 (2001)
17. Miller, K., Chinzei, K.: Mechanical properties of brain tissue in tension. J. Biomech. **35**(4), 483–490 (2002)
18. Miller, K., Chinzei, K.: Constitutive modelling of brain tissue: experiment and theory. J. Biomech. **30**(11–12), 1115–1121 (1997)
19. Fedorov, A., et al.: 3D Slicer as an image computing platform for the Quantitative Imaging Network. Magn. Reson. Imaging **30**(9), 1323–1341 (2012)
20. Claessens, M., Sauren, F., Wismans, J.: Modeling of the human head under impact conditions: a parametric study. In: SAE Transactions, pp. 3829–3848 (1997)
21. Miller, K.: Biomechanics of the Brain. Springer (2011)

22. Mazumder, M., et al.: Mechanical properties of the brain–skull interface. Acta Bioeng. Biomech. **15**(2), 3–11 (2013)
23. Famaey, N., et al.: Structural and mechanical characterisation of bridging veins: a review. J. Mech. Behav. Biomed. Mater. **41**, 222–240 (2015)
24. Walsh, D.R., et al.: Mechanical characterisation of the human dura mater, falx cerebri and superior sagittal sinus. Acta Biomater. **134**, 388–400 (2021)
25. Wittek, A., et al.: Patient-specific non-linear finite element modelling for predicting soft organ deformation in real-time; application to non-rigid neuroimage registration. Prog. Biophys. Mol. Biol. **103**(2–3), 292–303 (2010)
26. Miller, K., et al.: Beyond finite elements: a comprehensive, patient-specific neurosurgical simulation utilizing a meshless method. J. Biomech. **45**(15), 2698–2701 (2012)
27. Joldes, G.R., Wittek, A., Miller, K.: Suite of finite element algorithms for accurate computation of soft tissue deformation for surgical simulation. Med. Image Anal. **13**(6), 912–919 (2009)
28. Joldes, G.R., et al.: Realistic and efficient brain-skull interaction model for brain shift computation. In: Computational Biomechanics for Medicine III Workshop, MICCAI (2008)

HyperU-Mesh: Real-Time Deformation of Soft-Tissues Across Variable Patient-Specific Parameters

Sidaty El Hadramy[1,2(✉)], Nicolas Padoy[2,3], and Stéphane Cotin[1,2]

[1] Inria, Strasbourg, France
sidaty.el-hadramy@inria.fr
[2] ICube, University of Strasbourg, CNRS, Strasbourg, France
[3] IHU Strasbourg, Strasbourg, France

Abstract. Physics-based Patient-Specific Biomechanical models (PSBMs), particularly those using finite element methods (FEM), simulate organ behaviors accurately but are computationally intensive, especially for hyper-elastic tissues. To address this, U-Mesh [12] introduced a data-driven approach using U-Net architecture, achieving real-time inference but reliant on precise stiffness knowledge at training. This paper introduces HyperU-Mesh, an extension that integrates a Hypernetwork to condition U-Mesh based on stiffness prior distributions. By training with FEM-simulated data that varies stiffness under a predefined distribution, HyperU-Mesh ensures accuracy across variable stiffness without retraining. Experimental results highlight its effectiveness across different scenarios, showing comparable accuracy to FEM while significantly improving speed.

Keywords: Physics-based biomechanical models · Deep Learning · HyperNetworks

1 Introduction

In healthcare, digital twins (DTs) are virtual replicas of human organs and/or systems created using diverse medical imaging data [19]. Their application goes from training clinicians to intraoperative guidance. Several studies have demonstrated the benefits of physics-based digital twins [6,20], in particular, Patient-Specific Biomechanical models (PSBMs) for their accuracy and predictive ability over sparse and/or noisy intraoperative data. PSBMs are usually based on the principles of continuum mechanics, often employing finite element methods (FEM) to solve the numerical system describing the physical behavior of the organ or system. While these models provide precise simulations of organ and system behaviors, their computational complexity often results in time-consuming calculations, especially when involving soft tissue hyper-elastic behavior [11]. This limits their use in computer-assisted interventions, delaying

M. Kobielarz et al. (Eds.): MICCAI 2024, LNBE, pp. 129–139, 2025.
https://doi.org/10.1007/978-3-031-94128-3_12

real-time guidance during surgical procedures and complicating their effective integration. Therefore, significant efforts are being directed toward optimizing these models to enhance both the accuracy and computational speed of PSBMs, addressing these critical limitations [11].

Various trade-offs between the speed and accuracy of PSBMs have been proposed [6,12–14]. Haouchine et al. [6] suggested using a co-rotational model to handle large deformations with small strain. Yet, their approach significantly loses accuracy when advanced biomechanical laws are considered. Depending on the acceptable level of accuracy loss, reducing the model's degrees of freedom is a viable strategy to meet real-time constraints. Methods such as Proper Orthogonal Decomposition (POD) [13] and Proper Generalized Decomposition (PGD) [14] have been proposed for this purpose. Another category of methods leverages the high number of cores available in Graphics Processing Units (GPUs) for parallel computing, which enables significant speedups in handling computationally intensive problems [7]. Deep neural network architectures have recently demonstrated strong capabilities in learning complex, high-level nonlinear relationships between diverse input-output data [8,18]. One of the strengths of these networks is their ability to accurately perform inference in real-time when trained with sufficient data. Several works have proposed to train deep neural networks on simulated (using FEM) data [3,12,16,21], aiming to learn the behavior of PSBMs. Among them, U-Mesh, introduced by Mendizabal et al. [12], stands out as a simple yet effective solution.

U-Mesh is a data-driven approach based on a U-Net [17] architecture, designed to approximate the nonlinear relationship between forces and displacement fields. It is trained in a patient-specific manner using simulated data generated by Finite Element Method (FEM) and achieves real-time performance during inference. U-Mesh [12] has shown strong performance on real-world data, making it a promising approach in terms of both accuracy and speed. However, U-Mesh is trained on a single value of material stiffness. Hence, its accuracy at inference time depends on precise knowledge of patient-specific stiffness during training. This dependency limits its application in computer-assisted interventions, as such material parameters of organs are often available only during the intervention using techniques such as FibroScan®. This paper introduces HyperU-Mesh, an extended version of U-Mesh that incorporates conditioning on a distribution of material stiffness. This approach ensures that HyperU-Mesh maintains accuracy even when the stiffness is only known at inference time. Conditioning is achieved using a Hypernetwork [5], which generates a unique set of weights for U-Mesh based on the given stiffness value at inference time. HyperU-Mesh is trained on data simulated using the Finite Element Method (FEM). However, we generate samples where the material stiffness varies according to a specified prior distribution. The structure of the paper is as follows: Sect. 2 outlines our methodology. Section 3 illustrates its effectiveness in two distinct scenarios with varying tissue properties and domain geometries. Section 4 discusses our findings. Finally, Sect. 5 concludes the paper and explores future research directions.

2 HyperU-Mesh

Hypernetworks [5] are an approach where one network, referred to as the hypernetwork, generates the weights for another network, known as the primary network. This allows the primary network's weights to be dynamically adjusted during inference based on specific conditions inputted to the hypernetwork. Therefore, they have emerged as a way to enhance the flexibility and performance of deep neural networks [2]. We introduce HyperU-Mesh, an improved version of U-Mesh [12], incorporating a hypernetwork to integrate patient-specific stiffness information. An overview of HyperU-Mesh is illustrated in the Fig. 1. We use U-Mesh as a primary network that takes as input the forces, denoted F, and predicts the relative displacement field u. A 4-layer Multi-Layer Perceptron h predicts weights $d\theta$ for a given stiffness value λ.

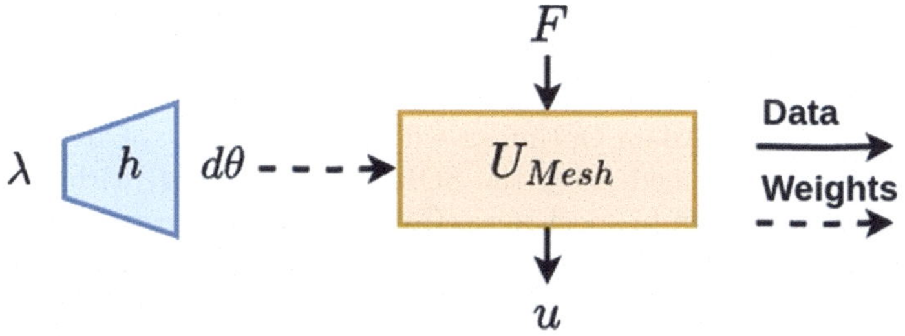

Fig. 1. Overview of HyperU-Mesh: U-Mesh [12] serves as the primary network, taking forces F as input to predict u, the relative displacement field. A hypernetwork h enhances U-Mesh's predictions by incorporating stiffness information. Both networks are trained end-to-end using data generated with FEM simulations.

Training hypernetworks is often challenging, particularly due to the proportionality between input and output magnitudes, which leads to very slow convergence. Ortiz et al. [15] identified and resolved this issue by treating the predicted weights as additive changes for the primary network. For a given training iteration n, the weights $d\theta_n$ predicted by the hypernetwork h are used to update the weights of the primary network (U-Mesh) following the equation below:

$$\theta_n = \theta_0 + d\theta_n \tag{1}$$

With θ_0 the initial weights of the U-Mesh. Unlike traditional hypernetworks [5], the weights θ_0 are also trainable parameters. The weights $d\theta_n$ influence the predictions of the U-Mesh by incorporating knowledge of the stiffness λ. This strategy permits a better initialization of the primary network weights, which leads to fast and efficient training of the U-Mesh. Typically, the training time of the approach is comparable to the training time of the primary network without a hypernetwork. h takes as input a parameter set λ that describes stiffness, varying depending on the specific biomechanical law used. For example, in the context of the Saint-Venant Kirchhoff constitutive law, λ corresponds to Young's modulus. The output of h has the same size as the number of parameters in the U-Mesh. We conducted an ablation study on the original version proposed by Mendizabal et al. [12] to decrease its weight count. As a result, without any loss of accuracy, we reduced the number of weights in the initial U-Mesh architecture by 4 by removing its unnecessary layers.

HyperU-Mesh is trained in an end-to-end manner using training data generated via FEM. Unlike U-Mesh, we generate training data by varying the stiffness parameter(s) according to a selected prior distribution from the literature. The training process optimizes the weights of both networks by minimizing the Mean Square Error (MSE) between the prediction and ground truth displacement fields.

3 Numerical Experiments

To evaluate our method, we conducted two experiments on two different geometries with distinct material properties. HyperU-Mesh is implemented with PyTorch[1]. We utilized Adam Optimizer with a learning rate of 10^{-5}. The FEM simulations were performed using the SOFA Framework [4] with the SOniCS [10] plugin for soft-tissue biomechanics. Computation was carried out using an Nvidia Titan RTX GPU. As a metric, we use classical Hausdorff distance in millimeters and the relative Hausdorff distance, defined as the percentage error relative to the deformation amplitude. Both metrics are computed using the implementation from SciPy[2]. We also report the computation time for a prediction.

3.1 PDMS Beam

We aim to estimate the deformation of a PolyDiMethylSiloxane (PDMS) beam under gravity, with the left end fixed to a vertical support, as illustrated in

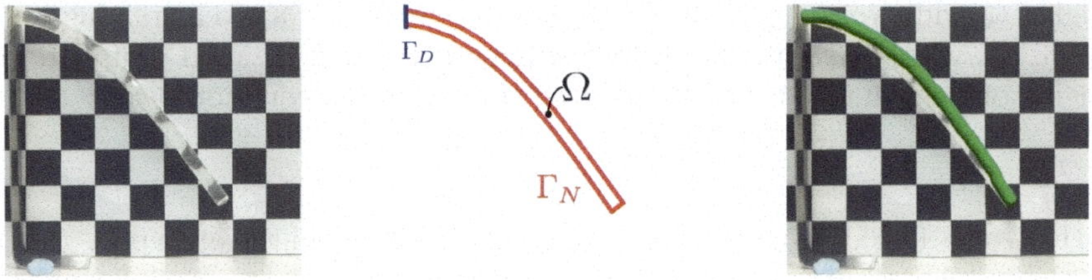

Fig. 2. Left: the observed deformation of the beam, fixed on the left side and deforming under gravity. **Middle**: An illustration of the simulation domain **Right**: The prediction of HyperU-Mesh (in green) overlaid onto the ground truth beam.

[1] https://pytorch.org/docs/stable/index.html.
[2] https://docs.scipy.org/doc/scipy/index.html.

the left image of Fig. 2. This example, adapted from Mazier et al. [9], involves characterizing the material behavior using the Mach-1^{TM} mechanical testing system (Biomomentum, Canada) for unconfined compression tests. The material is best approximated by a Mooney-Rivlin model with $C_{01} = 101$ kPa and $C_{10} = 151$ kPa. For validation, we obtained the ground truth deformation data from Mazier et al. [9].

Finite Element Simulation: We formulate a boundary value problem for computing the deformation of a hyperelastic material under both Dirichlet and Neumann boundary conditions (BCs). The beam occupies a volume Ω with boundary Γ. The Dirichlet and Neumann boundary conditions are on Γ_D and Γ_N, two subsets of Γ. These domains are illustrated in the middle image of Fig. 2. The elastic properties of soft tissues can be characterized using principles from continuum mechanics. By employing the Lagrangian formulation, the relationship between the deformed (x) and undeformed (X) states, at each point along the beam can be expressed as

$$\mathbf{x} = \mathbf{X} + \mathbf{u} \tag{2}$$

where u is the displacement field. The deformation gradient tensor $\boldsymbol{F} = \mathbf{I} + \nabla_{\mathbf{X}} \boldsymbol{u}$, provides a local description of the deformation and the Green-Lagrange strain tensor $\boldsymbol{E} \in \mathrm{R}^{3\times3}$ as expressed following Eq. 3

$$\boldsymbol{E} = \frac{1}{2}(\boldsymbol{C} - \boldsymbol{I}) \tag{3}$$

where $\boldsymbol{C} = \boldsymbol{F}^T \boldsymbol{F}$ is called the right Cauchy-Green deformation tensor and I the identity matrix. The material behaviour is approximated with a nearly incompressible Mooney-Rivlin [9] with parameters C_{01}, C_{10} and D_1, the strain-energy density function W is expressed as:

$$\mathbf{W} = \boldsymbol{C_{01}}(\boldsymbol{J}^{-\frac{2}{3}}\mathbf{I_C} - \mathbf{3}) + \boldsymbol{C_{10}}(\boldsymbol{J}^{-\frac{4}{3}}\mathbf{II_C} - \mathbf{3}) + \frac{\ln(\boldsymbol{J})}{\mathbf{2}\boldsymbol{D_1}} \tag{4}$$

In Eq. 4, $\mathbf{J}$ is the jacobian matrix, while $\mathbf{I_C} = \mathrm{tr}(\boldsymbol{C})$ and $\mathbf{II_C} = \frac{1}{2}\left((\mathrm{tr}(\mathbf{C}))^2 - \mathrm{tr}\left(\mathbf{C}^2\right)\right)$ are the classic invariants. The stress-strain relationship, also known as constitutive law, is obtained by differentiating $\mathbf{W}$ with respect to $\mathbf{C}$, see Eq. 5

$$\mathbf{S} = 2\frac{\partial \mathbf{W}}{\partial \mathbf{C}} \tag{5}$$

With S being the second Piola-Kirchhoff stress tensor. The boundary value problem is then formulated as in Eq. 6, where g represents the body forces, n the unit normal to Γ_N, and t the traction forces applied on Γ_N domain.

$$\begin{cases} \nabla(\boldsymbol{F}\boldsymbol{S}) = \boldsymbol{g} \quad \text{on} \quad \Omega \\ \boldsymbol{u}(\boldsymbol{X}) = \mathbf{0} \quad \text{on} \quad \Gamma_D \\ (\boldsymbol{F}\boldsymbol{S})\boldsymbol{n} = \boldsymbol{t} \quad \text{on} \quad \Gamma_N \end{cases} \tag{6}$$

The domain Ω is discretized wth hexahedral elements. This is motivated by their good convergence property, lock-free behavior and regular structure that fits well Convolutional Neural Networks (CNN) inputs [1]. Following a process similar to the U-Mesh data generation described by Mendizabal et al. [12], the weak formulation of Eq. 6 is solved using the finite element method (FEM). The implementation is carried out using the SOFA Framework [4]

Training Data Generation: We generated 6,000 training samples using the finite element simulation described earlier. For each sample, body forces were selected from a uniform distribution over [−15 N, −7 N], while the material parameters C_{01} and C_{10} were both sampled from uniform distributions over [100 kPa, 200 kPa]. The Dirichlet domain Γ_D remained fixed across all simulations. The relative displacement field for each dataset sample was obtained through the finite element simulation outlined earlier. A sample from the dataset includes the body forces F as input for U-Mesh, the relative displacement field u to supervise U-Mesh's prediction, and the material properties $\lambda = [C_{01}, C_{10}]$, which serve as the input for the hypernetwork h.

Results: HyperU-Mesh was trained during 100 epochs on the 6,000 generated samples, with a batch size of 1. We used an MSE between the prediction and the ground truth displacement fields as a loss function. Upon training, we tested HyperU-Mesh in the case where the body forces are equal to -9.81 N and the material parameters are: $C_{01} = 101$ kPa, $C_{10} = 151$ kPa and $D_1 = 7.965 \times 10^{-5}$ kPa. The predicted deformation (illustrated in the rightmost image of Fig. 2) is compared with the ground truth real deformation provided by Mazier et al. [9]. Table 1 presents the numerical results compared to the ground truth deformation from Mazier et al. [9], including the performance of U-Mesh [12], where all training was conducted with known material parameters. The table also compares the results of a FEM simulation to the ground truth deformation. The experiment demonstrates that HyperU-Mesh achieves comparable accuracy to both standard FEM and U-Mesh while operating 750 times faster than standard FEM. HyperU-Mesh also showcases versatility across a distribution of material

Table 1. Results from the beam experiments, shows comparisons between predictions made by standard FEM, U-Mesh, and HyperU-Mesh against the ground truth deformation. HyperU-Mesh achieves results comparable to both state-of-the-art methods while operating 750 times faster than standard FEM and offering greater flexibility than U-Mesh [12] in handling diverse material properties.

	Hausdorff (mm)	Relative Hausdorff (%)	Time (ms)
FEM	**4.7**	**3.6**	**3000**
U-Mesh [12]	**5.3**	**4.1**	**4**
HyperU-Mesh	**5.6**	**4.3**	**4**

parameters, unlike U-Mesh, which is constrained to a single known value required during training.

3.2 Human Liver

In this experiment, we aim to estimate the deformation of an ex-vivo human liver subjected to a force estimated to 5.1 N applied to the right lobe, the rest and deformed states of the liver are illustrated in Fig. 3. The geometries of both liver states are retrieved from a segmented CT scan. The stiffness of the organ was measured with a FibroScan® and estimated to be 7 kPa.

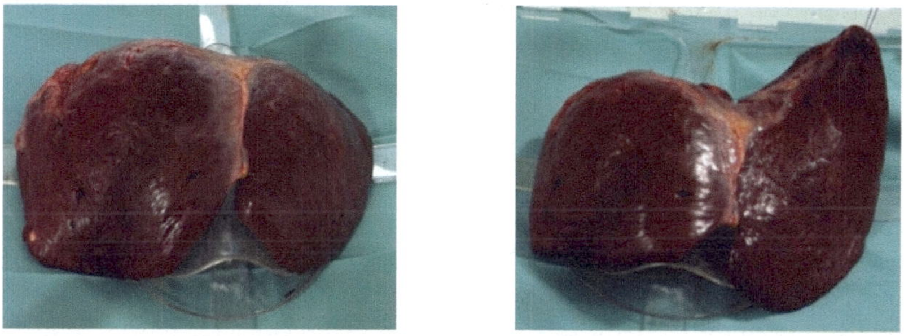

Fig. 3. Rest and deformed states of the ex-vivo liver.

Finite Element Simulation: Similar to the beam experiment, we formulate a boundary value problem to compute the deformation of soft tissue under both Dirichlet and Neumann boundary conditions. The underlying physics remains consistent with the beam case, except for the material behavior, which is approximated here using a Saint-Venant-Kirchhoff model. Thus, the strain energy density function is expressed as follows:

$$\mathbf{W} = \frac{Ev}{(1+v)(1-2v)}[\mathrm{tr}(\mathbf{E})]^2 + \frac{E}{2(1+v)}\mathrm{tr}(\mathbf{E}^2) \tag{7}$$

In Eq. 7, $\mathbf{E}$ is the Green-Lagrange strain tensor, while E and v being respectively the Young Modulus and the Poisson ratio. In Fig. 3, the Dirichlet domain is selected at the junction between the parenchyma and the portal vein. Traction forces are applied on the Neumann domain, which is chosen on the organ's surface.

Training Data Generation: Using finite element simulation, we generated 3,000 synthetic deformations of the organ. For each deformation, we randomly selected a location on the liver surface to apply traction forces, sampled from a uniform distribution over [2 N, 10 N]. Additionally, Young's modulus was sampled from a uniform distribution over [5000 Pa, 12000 Pa] for each generated sample. Consequently, a sample from the dataset includes the applied forces Fas input for U-Mesh, the chosen Young's modulus as the input λ of the hypernetwork h, and the relative displacement field obtained through FEM, which serves as the ground truth to supervise U-Mesh's prediction u.

Results: HyperU-Mesh was trained for 100 epochs on 3000 synthetic deformations of the liver with a batch size of 1. We used an MSE loss function between the predicted and FEM displacement fields to supervise the training. After training, we tested HyperU-Mesh with applied forces of 5.1N at the right lobe and a Young Modulus of 7 kPa. The predicted deformation was compared to the ground truth deformation (from the CT-scan), and the results are presented in Table 2. The leftmost image in Fig. 4 shows the geometry predicted by HyperU-Mesh superimposed on an image of the ground truth deformation. As one may notice, the errors in this experiment for HyperU-Mesh and other methods are higher than in the previous one. We attribute this to the uncertainty related to estimating the fixed boundary conditions. However, HyperU-Mesh achieves comparable results with U-Mesh and FEM, while being faster than FEM and more flexible (on material properties) than U-Mesh. To further evaluate the robustness of our method, we compared HyperU-Mesh predictions with standard FEM solutions over 100 synthetic deformations, resulting in a Mean Absolute Error (MAE) of 1.77 mm ± 0.82 mm. The rightmost image of Fig. 4 highlights a comparison between HyperU-Mesh and FEM predictions on a synthetic deformation.

Table 2. Results of the liver experiments compare the predictions of standard FEM, U-Mesh, and HyperU-Mesh against the ground truth deformation. HyperU-Mesh achieves results comparable to both state-of-the-art methods while being significantly times faster than standard FEM and more versatile than U-Mesh [12] in handling a range of material properties.

	Hausdorff (mm)	Relative Hausdorff (%)	Time (ms)
FEM	**15.1**	**11.6**	**500**
U-Mesh [12]	**16.6**	**12.7**	**4**
HyperU-Mesh	**16.5**	**12.6**	**4**

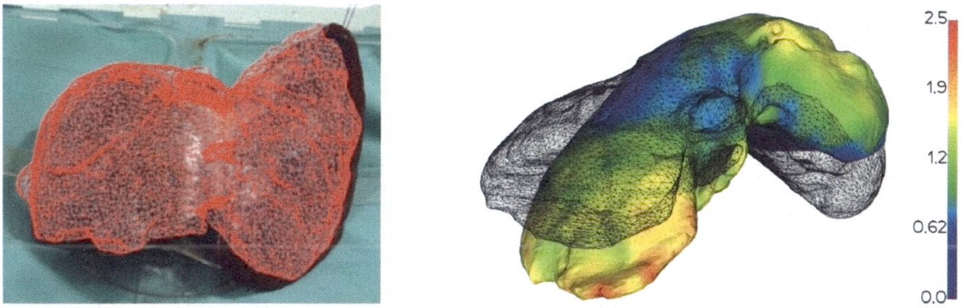

Fig. 4. Left: In the wireframe visualization, the prediction of HyperU-Mesh is shown when applying to the rest state (left image of Fig. 3) the same loads that produced the deformed state (right image of Fig. 3). The predicted result is superimposed over the ground truth deformation image for comparison. **Right**: The wireframe illustrates the liver's rest shape. The surface mesh represents a HyperU-Mesh prediction when specific loads are applied to this rest shape. The heatmap shows the errors of HyperU-Mesh compared to the FEM solution under the same loads.

4 Discussion

Results show that HyperU-Mesh is a promising approach, particularly in scenarios involving complex material properties and anatomical structures. The experiments demonstrated that HyperU-Mesh achieves accuracy comparable to both standard FEM and U-Mesh while operating significantly faster than standard FEM. Moreover, unlike U-Mesh, which is constrained to a single set of material parameters known during training, HyperU-Mesh exhibits versatility across a range of material parameters. This capability allows HyperU-Mesh to adapt to varying material properties without requiring retraining, thereby providing a more flexible and efficient solution for diverse simulation scenarios. This flexibility allows HyperU-Mesh to adapt to patient-specific material properties intraoperatively, enhancing its robustness and reliability in clinical applications where such variations are prevalent. However, the liver, being a complex organ with varying mechanical properties and physiological states, presents difficulties in accurately defining boundary conditions. The uncertainties associated with fixed boundary conditions can lead to increased errors across all modeling methods, as observed in experiment 2. Therefore, while HyperU-Mesh shows promise in enhancing the efficiency and accuracy of biomechanical simulations, particularly in liver modeling, several challenges remain, including handling uncertain boundary conditions. Addressing these challenges will be crucial in realizing the full potential of HyperU-Mesh in clinical practice.

5 Conclusion

In this paper, we introduced HyperU-Mesh, an extension of the U-Mesh [12] framework designed to enhance the accuracy, speed, and flexibility of physics-based biomechanical models (PSBMs) in real-time applications. PSBMs, particularly those based on finite element methods (FEM), are accurate for simulating organ behaviors, yet their computational intensity has hindered real-time implementation, especially for hyper-elastic tissues. HyperU-Mesh addresses these challenges by integrating Hypernetwork into the U-Mesh architecture. This allows HyperU-Mesh to adapt dynamically to patient-specific biomechanical properties, such as varying stiffness parameters, without the need for retraining. Thus, HyperU-Mesh ensures robust performance in real-world scenarios where precise material parameters are often only available intraoperatively. Our experimental results demonstrated the effectiveness of HyperU-Mesh across different scenarios involving diverse complex material properties and geometries. Highlighting its ability to predict deformations with accuracy comparable to traditional FEM simulations while being real-time at inference. In future work, we will extend the application of HyperU-Mesh to include fixed boundary conditions and geometric variation uncertainties.

References

1. Benzley, S.E., et al.: A comparison of all hexagonal and all tetrahedral finite element meshes for elastic and elasto-plastic analysis (2011)
2. Chauhan, V.H., et al.: A brief review of hypernetworks in deep learning (2023)
3. El Hadramy, S., et al.: Towards real-time vessel guided augmented reality for liver surgery. In: 2024 IEEE International Symposium on Biomedical Imaging (ISBI) (2024)
4. Faure, F., et al.: SOFA: a multi-model framework for interactive physical simulation. In: Studies in Mechanobiology, Tissue Engineering and Biomaterials, pp. 283–321. Springer, Berlin, Heidelberg (2012)
5. Ha, D., Dai, A., Le, Q.V.: HyperNetworks (2016)
6. Haouchine, N., et al.: Image-guided simulation of heterogeneous tissue deformation for augmented reality during hepatic surgery. In: 2013 IEEE International Symposium on Mixed and Augmented Reality (ISMAR). IEEE, Adelaide, Australia (2013)
7. Johnsen, S.F., et al.: NiftySim: a GPU-based nonlinear finite element package for simulation of soft tissue biomechanics. Int. J. Comput. Assist. Radiol. Surg. **10**(7), 1077–1095 (2015)
8. LeCun, Y., Bengio, Y.: Convolutional networks for images, speech, and time series. In: The Handbook of Brain Theory and Neural Networks, pp. 255–258. MIT Press, London, England (1998)
9. Mazier, A., et al.: Inverse deformation analysis: an experimental and numerical assessment using the FEniCS Project. Eng. Comput. **38**(5), 4099–4113 (2022)
10. Mazier, A., et al.: Sonics: develop intuition on biomechanical systems through interactive error controlled simulations. Eng. Comput. **40**(3), 1857–1876 (2024)
11. Meier, U., et al.: Real-time deformable models for surgery simulation: a survey. Comput. Methods Prog. Biomed. **77**(3), 183–197 (2005)
12. Mendizabal, A., Marquez-Neila, P., Cotin, S.: Simulation of hyperelastic materials in real-time using deep learning. Med. Image Anal. **59**, 101569 (2020)
13. Niroomandi, S., et al.: Real-time deformable models of non-linear tissues by model reduction techniques. Comput. Methods Prog. Biomed. **91**(3), 223–231 (2008)

14. Niroomandi, S., et al.: Real-time simulation of biological soft tissues: a PGD approach. Int. J. Numer. Method. Biomed. Eng. **29**(5), 586–600 (2013)
15. Ortiz, J.J.G., Guttag, J., Dalca, A.: Magnitude invariant parametrizations improve hypernetwork learning (2023)
16. Roewer-Despres, F., Khan, N., Stavness, I.: Towards finite-element simulation using deep learning (2018)
17. Ronneberger, O., Fischer, P., Brox, T.: U-Net: convolutional networks for biomedical image segmentation. In: Navab, N., Hornegger, J., Wells, W.M., Frangi, A.F. (eds.) MICCAI 2015. LNCS, vol. 9351, pp. 234–241. Springer, Cham (2015). https://doi.org/10.1007/978-3-319-24574-4_28
18. Scarselli, F., et al.: The graph neural network model. IEEE Trans. Neural Netw. **20**(1), 61–80 (2009)
19. Servin, F., et al.: Simulation of image-guided microwave ablation therapy using a digital twin computational model. IEEE Open J. Eng. Med. Biol. **5**, 107–124 (2024)
20. Suwelack, S., et al.: Physics-based shape matching for intraoperative image guidance. Med. Phys. **41**(11), 111901 (2014)
21. Tonutti, M., Gras, G., Yang, G.Z.: A machine learning approach for real-time modelling of tissue deformation in image-guided neurosurgery. Artif. Intell. Med. **80**, 39–47 (2017)

Combined Method for the Assessment of Occlusal Loadings Based on the Digital Measurements

Grzegorz Milewski(✉), Agnieszka Chojnacka-Brożek, Weronika Idzik, and Sylwia Majcher

Department of Applied Mechanics and Biomechanics, Cracow University of Technology, Cracow, Poland
grzegorz.milewski@pk.edu.pl

Abstract. Occlusion refers to the mutual contact between opposing tooth surfaces in the dental arches. The action of the jaw elevator muscles produces the bite force which may vary widely in magnitude and direction due to the individual anatomy, jaw biomechanics and reflex mechanisms. Articulation foil or tracing paper are the most commonly used method in clinical practice for assessing dental arch alignment. These diagnostic methods are able to determine the exact location of occlusal contacts, but they are not able to measure the bite forces themselves. Currently there are several methods for measuring occlusal forces, although they are not widely used. These methods include mechanical occlusal force gauges measurements and digital scanning approaches. The first method provides the maximum force during biting for a separate tooth, while the second method estimates only the qualitative percentage distribution of occlusal loads in the individual sections of the dental arches. The aim of the work is to propose a hybrid estimation of occlusal forces for incisors, canines and lateral teeth (both premolars and molars) of upper and lower jaws based on both occlusal force gauges measurements and digital scanning. These combined method provides quantitative estimation of bite distribution in all individual tooth groups of the lower and upper jaws, which is of extreme importance for biomedical engineers, especially in the case of numerical analyzes by means of the finite element method.

Keywords: occlusal loadings · digital measurements · hybrid bite assessment

1 Introduction

The determination of bite force is considered in a contemporary dentistry as a crucial procedure in assessing the features of individual patient masticatory system concerning its deformities and pathologies, such as malocclusion or over loading. It is also an important parameter to detect premature tooth contacts, quality of bite for adult patients as well as function and efficacy of dental implants, prostheses and orthodontic treatments. Traditionally, occlusal contacts have been detected using articulating paper or foil. Although these diagnostic methods are able to determine the exact location of occlusal contacts,

M. Kobielarz et al. (Eds.): MICCAI 2024, LNBE, pp. 140–151, 2025.
https://doi.org/10.1007/978-3-031-94128-3_13

they are not able to measure their forces reliably. At present, many devices are used to record the bite force. Most of these devices use force transducers to detect bite force, such as strain gauge transducers, piezoresistive transducers, piezoelectric transducers, optical fiber transducers, and pressure-sensitive films [1, 2]. Bite forces in adults are highly variable. Generally, it is assumed that in the area of the front teeth, the maximum occlusal forces average between 200 and 300 N, while in the area of the molars and premolars, they range from 700 to 800 N [3, 4, 5, 6]. In individual cases, these values can, however, be significantly exceeded [7, 8].

The application of digital technologies in dentistry has advanced rapidly in the past 2 -3 decades due to the development of CAD/CAM (computer-aided design and computer-aided manufacturing) technologies including 3D software and imaging procedures, digital sensors, digital tensometry, 3D dental scanner constructions. The scanners produce datasets with varying quality based on different scanning technologies and software processing algorithms. These digital technologies are crucial tools in diagnosing, treatment planning, and monitoring progress in orthodontic or prosthetic treatments, enabling a more precise and personalized approach to occlusion issues in patients [9]. The fundamental goal of digital dentistry is to evaluate precisely the surface topography of maxillary and mandibular arches, teeth contoured surfaces, as well implants and prostheses in the anterior, posterior position and interproximal areas [10, 11]. Digital occlusal analyzers expand the possibilities for determining the quality of bite [12, 13] and can be used to measure the right- and left-side balance of occlusal contact forces in the detection and correction of temporomandibular disorders [14] or parafunctional and traumatic conditions of whole-mouth occlusions including the resolution of dental problems such as bruxism, severe tooth wear, and muscle pain [15]. An increasing interest in the application of the digital devices can be observed in the evaluation of implant and prostheses application [16–20] and in the assessment of orthodontic treatment [21, 22].

The idea of the combined/hybrid method stems from the fact that traditionally used analog methods measure the total/absolute bite force accumulated at the location of the measurement sensor (gauge transducers). On the other hand, digital methods provide only a qualitative (percentage) distribution of occlusion in the individual dental sections of both jaws, distinguishing between the mandible and maxilla, as well as the left/right balance.

The main contribution/aim of the work is to present an algorithm for determining the distribution and values of local bite forces in individual dental groups (incisors, canines, and molars) for the mandible and maxilla using the digital T-scan method, based on a priori values of absolute bite force measured point by point using the analog method in specific dental sections, as reported in the literature. The inclusion criterion applied in the algorithm was a comparable age group of patients.

2 Methods

The study was conducted on a group of 20 individuals, consisting of 10 males and 10 females, aged between 20 and 26 years using the T-scan® Novus™ digital measuring system by Tecscan INC [23]. The T-scan device assists dentists in occlusion diagnosis during patient bite analysis. It is used to detect premature tooth contacts, cumulative occlusion, bite irregularities and finally provides the percentage distribution of loads in the individual sections of the dental arches.

All the patients had permanent and natural teeth. The quality of the bite includes both normal and pathological conditions. Most of the subjects had undergone numerous corrections of their teeth using composite fillings. In the research group 50% of women and 10% of men had previously corrected their bite with orthodontic appliances and currently stabilized the anterior section of the teeth with a permanent retainer. In the group of 10 examined women, 8 out of 10 cases concerned the predominance of occlusal forces on the molars. The remaining two cases are women whose teeth are characterized by a very symmetrical distribution of forces on the dental arches. In the group of 10 studied men, 5 for10 cases concern the predominance of occlusal forces on the molars. The remaining cases are men whose teeth are characterized by a symmetrical distribution of forces on the dental arches. Additionally, there was one case in which occlusal forces were accumulated on the incisors. Occlusion distribution and its symmetry in the study group are shown in Fig. 1.

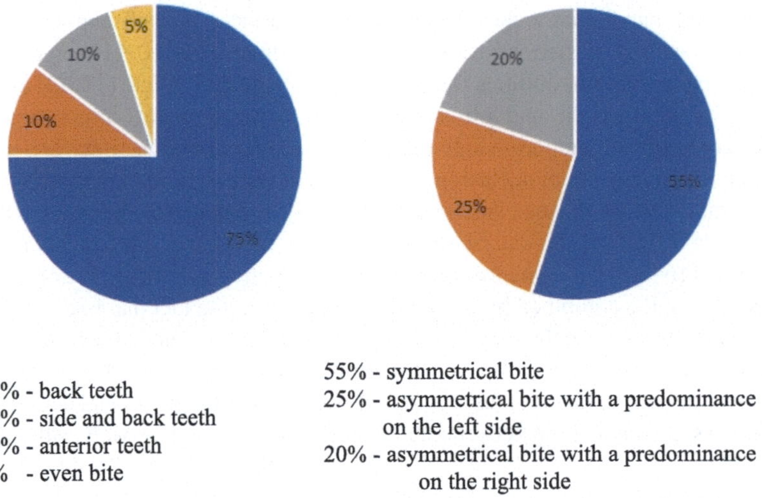

Fig. 1. Occlusion distribution and its symmetry in the study group.

Individual teeth bite forces are available in the literature [3–6]. Due to the comparable study group the reference maximum bite forces were taken after the data provided in the publication [5] in the list of cited literature - Table 1. This paper presents the results of absolute/maximal bite force studies for a group of 163 young adult patients aged 20 to 30 years what corresponds to the research group examined in our studies with the use of the T-scan digital method.

Table 1. The average values of maximum bite forces for adult females and males are given in Newtons [5].

Teeth group	Incisors	Canines	Lateral Teeth
Women	185,5	325,3	365,1
Men	226,5	401,5	524,4

The proposed algorithm for the hybrid method of assessing occlusal forces in the individual sections of the lower and upper dental arches is based on the following steps:

1. digital measurement of the percentage distribution of the patient's occlusion in the individual sections of the upper and lower dental arches,
2. incorporation in the algorithm, based on the literature, the dominant absolute bite value as the reference input parameter value in the proposed algorithm; this value depends on the type of the patient's bite (even distribution of occlusion, retrognathism, prognathism),
3. final assessment of individual occlusal forces for the three main sections of teeth, i.e. incisors, canines, and posterior teeth of both dental arches, based on the percentage distribution of the total cumulative occlusion.

3 Results and Discussion

Figures 2, 3, 4, 5, 6 and 7 show respectively the results of digital scan measurements for the chosen representative patients, both women (W) and men (M), for variant differentiated occlusal loading systems with the variant bite forces concentration in the lateral and anterior tooth sections, for even distribution of occlusion in the dental arches, symmetrical and asymmetrical bite distributions.

Patient W1 has undergone orthodontic treatment and was characterized by even distribution of occlusion forces between the left and right parts of the dental arch: 49.7% and 50.3%. The force concentrated on the molars is 62.4% on the upper arch and 73.4% on the lower arch.

Patient W7 had a full permanent dentition. Due to the nature of the tooth setting, the test results did not show the exact outline of the arches on the molars. The patient has not undergone orthodontic treatment. The study shows a large discrepancy in the distribution of occlusion forces from left to right: 70.9% and 20.1%. The left molars are subjected to significant load. The greatest accumulation of forces occurs on teeth 27 and 37.

Patient W10 had a full permanent dentition and was characterized by a very even and symmetrical distribution of occlusal forces throughout both dental arches. The patient has undergone orthodontic treatment. The distribution of occlusion forces between the left and right sides of the arch is 56.5% and 43.5%.

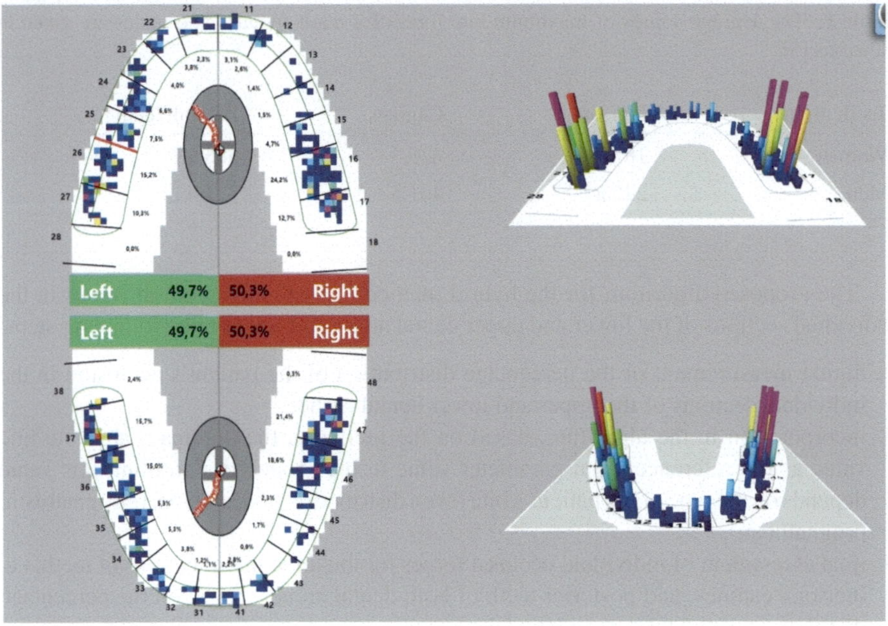

Fig. 2. The occlusion analysis result of patient W1 with the use of T-scan digital system.

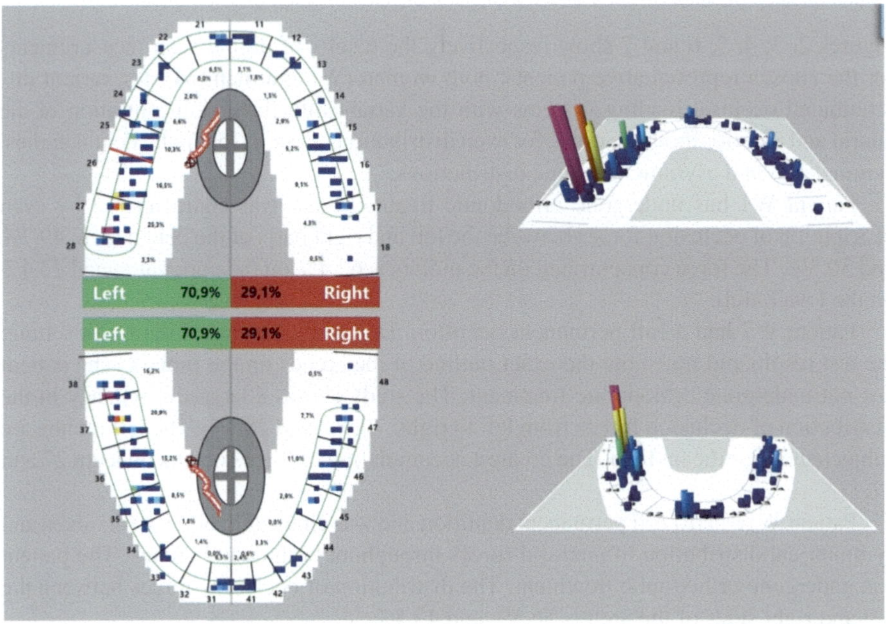

Fig. 3. The occlusion analysis result of patient W7 with the use of T-scan digital system.

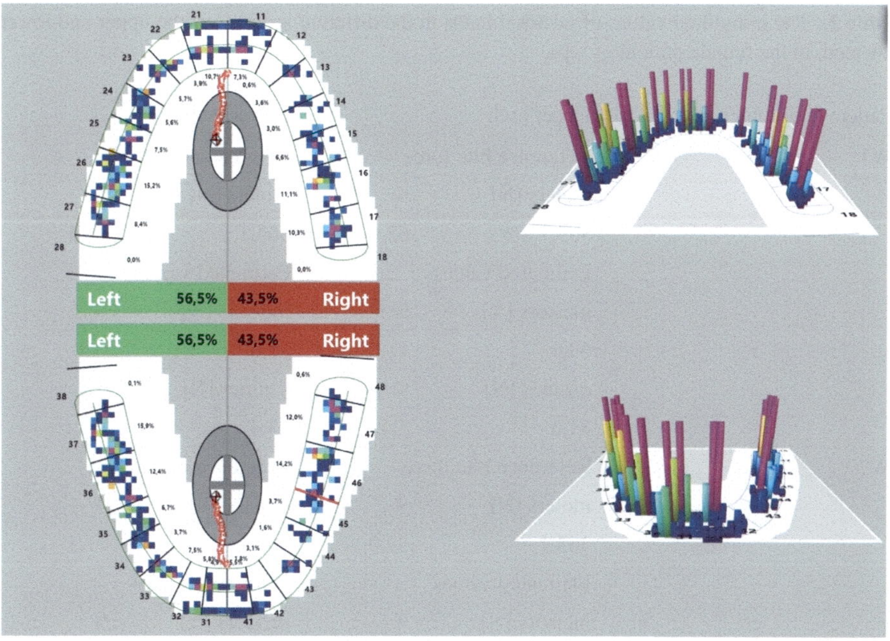

Fig. 4. The occlusion analysis result of patient W10 with the use of T-scan digital system.

Determination of the individual bite force values, based on the algorithm proposed in the paper, in all tooth groups of upper and lower jaws for the presented in the paper female cases is shown in Table 2.

Patient M1 had a full permanent dentition. The patient did not undergo orthodontic treatment. The maximum forces are concentrated on molars and premolars. The distribution of forces from left to right is 55.1% and 44.9%.

Patient M4 had a full permanent dentition. The distribution of forces between the left and right sides of the arches is very asymmetric: 30.2% and 69.8%. The maximum forces are generated on the molars, especially around the teeth 16 and 17. One can also notice small values in the lower arch on the incisors, what is due to the fact that no contact was detected on teeth 31 and 41.

Patient M10 had a full permanent teeth with a bite focused on the anterior part of the arch. Patient has not undergone orthodontic treatment. The occlusion defect is significant and may have a very negative impact on the functioning of the entire stomatognathic system. The distribution of forces between the left and right sides is 43.4% and 56.6%.

Table 2. The individual values of occlusal loads in the different sections of the upper and lower jaw teeth in the female patient group.

Patient no (women – W)	Upper arch		Lower arch	
W1	Reference bite force		Reference bite force	
	molars [N]	%	molars [N]	%
	365,1	62,4	365,1	73,4
	Estimated values		Estimated values	
	incisors [N]	%	incisors [N]	%
	69,0	11,8	31,3	6,3
	canines [N]	%	canines [N]	%
	31,6	5,4	23,4	4,7
W7	Reference bite force		Reference bite force	
	molars [N]	%	molars [N]	%
	365,1	59,0	365,1	71,5
	Estimated values		Estimated values	
	incisors [N]	%	incisors [N]	%
	70,5	11,4	48,5	9,5
	canines [N]	%	canines [N]	%
	21,7	3,5	24,0	4,7
W10	Reference bite force		Reference bite force	
	incisors [N]	%	incisors [N]	%
	185,5	22,5	185,5	18,0
	Estimated values		Estimated values	
	canines [N]	%	canines [N]	%
	76,7	9,3	109,2	10,6
	molars [N]	%	molars [N]	%
	371,0	45,0	568,9	55,2
	Reference bite force		Reference bite force	
	molars [N]	%	molars [N]	%
	365,1	45,0	365,1	55,2
	Estimated values		Estimated values	
	incisors [N]	%	incisors [N]	%
	182,6	22,5	119,1	18,0
	canines [N]	%	canines [N]	%
	75,5	9,3	70,1	10,6

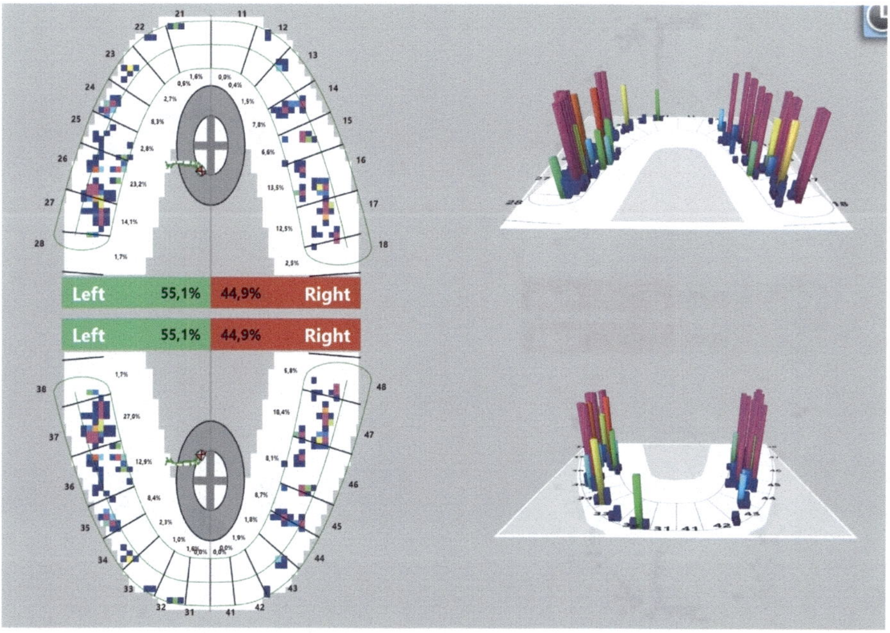

Fig. 5. The occlusion analysis results of patient M1 with the use of T-scan digital system.

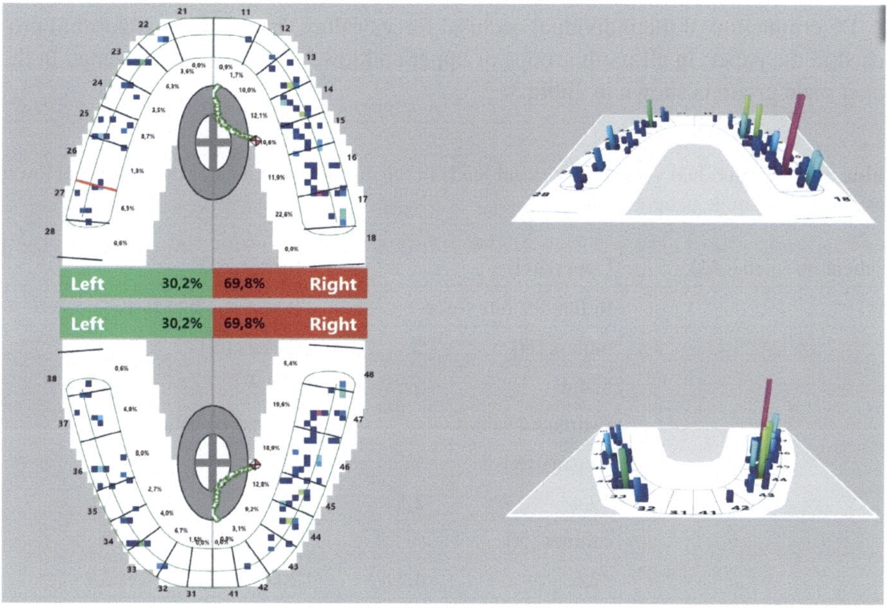

Fig. 6. The occlusion analysis results of patient M4 with the use of T-scan digital system.

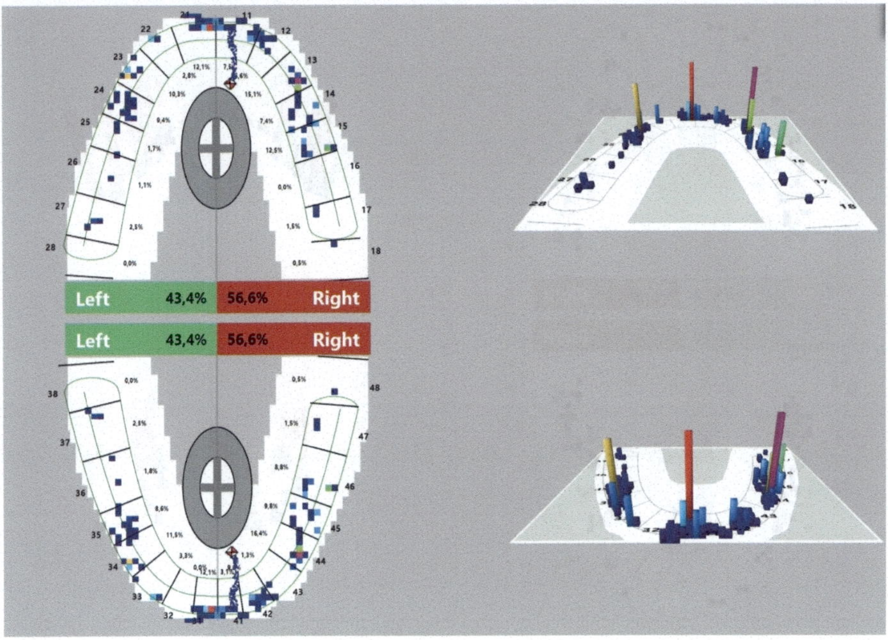

Fig. 7. The occlusion analysis results of patient M10 with the use of T-scan digital system.

Determination of the individual occlusal forces values, based on the algorithm proposed in the paper, in all tooth groups of upper and lower jaws for the presented in the paper male group is shown in Table 3.

Table 3. The individual values of occlusal loads in the different sections of the upper and lower jaw teeth in the male patient group.

Patient no (men – M)	Upper arch		Lower arch	
M1	Reference bite force		Reference bite force	
	molars [N]	%	molars [N]	%
	524,4	67,5	524,4	73,9
	Estimated values		Estimated values	
	incisors [N]	%	incisors [N]	%
	19,4	2,5	11,4	1,6
	canines [N]	%	canines [N]	%
	32,6	4,2	20,6	2,9

(continued)

Table 3. (*continued*)

Patient no (men – M)	Upper arch		Lower arch	
M4	Reference bite force		Reference bite force	
	molars [N]	%	molars [N]	%
	524,4	42,7	524,4	59,3
	Estimated values		Estimated values	
	incisors [N]	%	incisors [N]	%
	76,1	6,2	21,2	2,4
	canines [N]	%	canines [N]	%
	200,2	16,3	86,7	9,8
M10	Reference bite force		Reference bite force	
	incisors [N]	%	incisors [N]	%
	226,5	39,0	226,5	25,0
	Estimated values		Estimated values	
	canines [N]	%	canines [N]	%
	147,5	25,4	41,7	4,6
	molars [N]	%	molars [N]	%
	32,5	5,6	136,8	15,1

4 Conclusions

The methodology for estimating occlusal forces in individual sections of dental arches has been proposed due to the lack of clear quantitative data in the assessment of total occlusion with the digital T-scan device. The hybrid method of bite estimation is an extension of the up-to-date used methods for the individual assessment of occlusion.

The research indicates that the highest bite forces occur in the molar teeth in the lower arch, both for women and men. Comparison of the calculated values of occlusal forces in the anterior parts of both jaws proves that those in the upper jaw are higher. For the majority of women, with one exception for the lower arch, significantly higher values were noted in the incisor areas than for canines. On the other hand, in the group of men, the obtained results revealed opposite regularity in the upper arch, where the canine section was generally more stressed than the incisor one.

The method presented in the paper is effective, relatively simple, and produces the results which are reliable and validated in numerous experimental studies.

References

1. Gu, Y., Bai, Y., Xie, X.: Bite force transducers and measurement devices. Front. Bioeng. Biotechnol. **9**, 665081 (2021). https://doi.org/10.3389/fbioe.2021.665081

2. van Vuuren, L.J., van Vuuren, W.A.J., Broadbent, J.M., et al.: Development of a bite force transducer for measuring maximum voluntary bite forces between individual opposing tooth surfaces. J. Mech. Behav. Biomed. Mater. **109**, 103846 (2020). https://doi.org/10.1016/j.jmbbm.2020.103846
3. Proffit, W.B., Fields, H.W., Nixon, W.L.: Occlusal forces in normal and long face adults. J. Dent. Res. **5**, 566–571 (1983)
4. Korioth, T.W.P., Waldron, T.W., Versluis, A., et al.: Forces and moments generated at the dental incisors during forceful biting in humans. J. Biomech. **6**, 631–633 (1997)
5. Chladek, W., Lipski, T., Karasinski, A.: Experimental evaluation of occlusal forces. Acta Bioeng. Biomech. **3**, 25–37 (2001)
6. Varga, S., Spalj, S., Varga, M.L., et al.: Maximum voluntary molar bite force in subjects with normal occlusion. Eur. J. Orthodontics **33**, 427–433 (2011). https://doi.org/10.1093/ejo/cjq097
7. Osborne, J.W.: Features of human jaw design which maximize the bite force. J. Biomech. **5**, 589–595 (1996)
8. Milewski, G.: Numerical and experimental analysis of effort of human tooth hard tissues in terms of proper occlusal loadings. Acta Bioeng. Biomech. **1**, 47–58 (2005)
9. Chowdhary, R., Sonnahalli, N.K.: Clinical applications of the T-Scan quantitative digital occlusal analysis technology: a systematic review. Int. J. Comput. Dent. **27**(1), 49–86 (2024). https://doi.org/10.3290/j.ijcd.b3945153
10. Lee, S.J., Kim, S.-W., Lee, J.J., Cheong, C.W.: Comparison of intraoral and extraoral digital scanners: evaluation of surface topography and precision. Dent. J. (Basel). **8**(2), 52, 20 May 2020. https://doi.org/10.3390/dj8020052
11. Fraile, C., Ferreiroa, A., Rubio, M.R., et al.: Clinical study comparing the accuracy of interocclusal records, digitally obtained by three different devices. Clin. Oral Investig. **26**, 4663–4668 (2022). https://doi.org/10.1007/s00784-022-04542-6
12. Levartovsky, S., Peleg, G., Matalon, S., et al.: Maximal bite force measured via digital bite force transducer in subjects with or without dental implants—a pilot study. Appl. Sci. **12**, 1544 (2022). https://doi.org/10.3390/app12031544
13. Jauregi, M., Amezua, X., Mikel Iturrate, M., et al.: Improving the precision of recordings acquired with digital occlusal analyzers: a dental technique. J Prosthet Dentist **132**, 37–41 (2024). https://doi.org/10.1016/j.prosdent.2023.08.001
14. Jauregi, M., Amezua, X., Iturrate, M., et al.: Repeatability and reproducibility of 2 digital occlusal analyzers for measuring the right - and left-side balance of occlusal contact forces: an in vitro study. J. Prosthet. Dent. **132**, 179–187 (2024). https://doi.org/10.1016/j.prosdent.2023.07.026
15. Ortún-Terrazas, J., Cegoñino, J., Amaya Pérez del Palomar, A.: In silico study of cuspid' periodontal ligament damage under parafunctional and traumatic conditions of whole-mouth occlusions. A patient-specific evaluation. Comput. Methods Programs Biomed. **184** (2020). https://doi.org/10.1016/j.cmpb.2019.105107
16. Lerner, H., Mangano, C., Luongo, G., et al.: Implant complications after installation with traditional vs digital occlusal indicators. Adv. Dent Tech. **13**, 1–15 (2024)
17. Gassara, Y., Kallala, R., Nasri, S., et al.: Digital occlusal analyses in fixed prosthodontics. Int. Dent. J. **74**, 227–228 (2024). https://doi.org/10.1016/j.identj.2024.07.078
18. Zakkour. J.D., Zakkour, D.D., Javier Montero, J., et al.: Digital analysis of occlusion in fixed partial implant prostheses: how to overcome age-related changes in the stomatognathic system. Prosthesis **6**, 119–134 (2024). https://doi.org/10.3390/prosthesis6010010
19. Luo, Q., Ding, Q., Zhang, L., et al.: Analyzing the occlusion variation of single posterior implant - supported fixed prostheses by using the T-scan system: a prospective 3-year follow-up study. J. Prosthet. Dent. **123**, 79–84 (2020). https://doi.org/10.1016/j.prosdent.2018.12.012

20. Aradya, A., Nagarajagowd, R.S.K., Basavarajm, R.M., et al.: Influence of T-scan system on occlusion correction of implant supported prostheses: a systematic review. J. Contemp. Dent. Pract. **23**(1), 105–117 (2022). https://doi.org/10.5005/jp-journals-10024-3288
21. Wang, M., Liu, L., Ma, X., et al.: Computerized dynamic occlusal analysis and its correlation with static characters in post-orthodontic patients using the T-Scan system and the ABO objective grading system. BMC Oral Health **23**, 312 (2023). https://doi.org/10.1186/s12903-023-02868-5
22. Milewski, G., Hille, A.: Experimental strength analysis of orthodontic extrusion of human anterior teeth. Acta Bioeng. Biomech. **1**, 15–21 (2012)
23. Tekscan: T-scan 9.1 Computerized Occlusal Analysis System, Tekscan, INC, Boston (2015)

Author Index

M. Kobielarz et al. (Eds.): MICCAI 2024, LNBE, pp. 153–154, 2025.
https://doi.org/10.1007/978-3-031-94128-3

The manufacturer's authorised representative in the EU is Springer Nature Customer Service Centre GmbH, Europaplatz 3, 69115 Heidelberg, Germany. If you have any concerns regarding our products, please contact ProductSafety@springernature.com

Printed and bound by CPI Group (UK) Ltd, Croydon, CR0 4YY
30/09/2026
02225577-0010